PRINCIPLES OF BUILDING ECONOMICS

AN INTRODUCTION

John Raftery

Reader in Construction Economics
Thames Polytechnic
School of Surveying

OXFORD
BSP PROFESSIONAL BOOKS
LONDON EDINBURGH BOSTON
MELBOURNE PARIS BERLIN VIENNA

Chapter 11 is a revised version of a paper first published in 1987 under the title 'The state of cost/price modelling in the UK construction industry: a multicriteria approach.' In Brandon, P. ed (1987) Building Cost Models and Computers. London: E&FN Spon.

BSP Professional Books
A division of Blackwell Scientific Publications Ltd
Editorial offices:
Osney Mead, Oxford OX2 0EL
25 John Street, London WC1N 2BL
23 Ainslie Place, Edinburgh EH3 6AJ
3 Cambridge Center, Cambridge, MA 02142, USA
54 University Street, Carlton Victoria 3053, Australia

Other Editorial Offices:
Librairie Arnette SA
2, rue Casimir-Delavigne
75006 Paris
France

Blackwell Wissenschafts-Verlag
Meinekestrasse 4
D-1000 Berlin 15
Germany

Blackwell MZV
Feldgasse 13
A-1238 Wien
Austria

First published 1991
Reprinted 1992

Set by Setrite Typesetters
Printed and bound in Great Britain by
Hartnolls, Bodmin, Cornwall

DISTRIBUTORS

Marston Book Services Ltd
PO Box 87
Oxford OX2 0DT
(*Orders:* Tel: 0865 791155
Fax: 0865 791927
Telex: 837515)

USA
Blackwell Scientific Publications, Inc.
3 Cambridge Center
Cambridge, MA 02142
(*Orders:* Tel: 800 759-6102
617 225-0401)

Canada
Oxford University Press
70 Wynford Drive
Don Mills
Ontario M3C 1J9
(*Orders:* Tel: 416 441-2941)

Australia
Blackwell Scientific Publications (Australia) Pty Ltd
54 University Street
Carlton, Victoria 3053
(*Orders:* Tel: 03 347-0300)

British Library
Cataloguing in Publication Data

Raftery, John
Principles of building economics: an introduction
1. Buildings. Construction. Costing
I. Title
690.0681

ISBN 0-632-02917-X

To my parents

Contents

Preface

The aim of this book is to provide a critical introduction to some important and fundamental aspects of building economics. The book is intended to be a core text for students and it is hoped it will be read by practising professionals wanting a stimulating and at times provocative 'read'. It draws on ideas from economics to describe and understand the production of the built environment.

The need for a book such as this first became clear to me when I began to teach courses on building economics to students reading for degrees in architecture, engineering and surveying. I quickly learned that it was just not possible to write short reading lists as the necessary material was scattered across a wide range of books and journals. While it was possible to recommend standard texts on the *techniques* of building economics there was no one, authoritative but accessible, source on the underlying principles. This is an important gap in the literature, as in volatile economic conditions the use of techniques needs to be 'intelligent'. That is to say, the user needs to be aware of how changes in the economic environment affect the inputs and outputs of plans and forecasts.

This book is, therefore, intended to complement and be used in conjuction with a text on techniques such as Ferry and Brandon (1991). Together they would provide a rigorous and comprehensive introduction to the theory and practice of building economics.

The book is in three parts. The first part contains a macro and micro economic introduction to the subject which considers how economic thought has developed since Adam Smith and goes on to introduce key issues such as the current divergence of opinion in economics, markets, price determination and the role of government in a mixed economy. The fundamental concepts of cost, price and value are considered in detail. The text takes account of how 'green accounting' produces an entirely different definition of cost.

Part two considers the demand side beginning with a consideration of the problems of intertemporal choice in construction project appraisal. There are chapters on demand for construction and on the problem of obsolescence in the built environment.

Part three is an examination of the supply side of the industry. This part contains chapters on the theory of the firm, the operation of the firm, price determination for construction projects, the design process and cost models.

The eclectic nature of a text such as this makes it impossible to give more than the briefest of introductions to each subject. An extensive bibliography is provided for those who wish to pursue areas of individual interest. Throughout the text the use of statistics and mathematics has been kept to a minimum.

John Raftery
Greenwich
May 1990

Acknowledgements

I owe a great many intellectual debts to my colleagues, teachers and students. I am especially grateful to Henryk Hajduk and Marek Bryx for offering me the opportunity to present some ideas on bidding, markets and price determination at The Central School of Planning and Statistics in Warsaw. I have benefited from many discussions with colleagues from the Working Commission on Building Economics of the International Council for Building Research and Documentation (CIB), in particular, Harold Marshall, Olli Niemi, Raimo Salokangas, Klara Szoke and Alan Wilson. Many of the ideas in this text were discussed and sharpened in the economics seminars of the MA in Building Rehabilitation Studies in the year 1988/9. I am grateful to Paul Hodgkinson and Sue Lee for producing the drawings and to Jenny Lynch for word processing.

I am indebted to Brian Atkin, Paul Balchin, Ranko Bon, John Connaughton, David Jaggar, Stephen Mak, David Morely, Sidney Newton, Martin Skitmore, Stan Smith, and David Wills for specific comments, criticisms and suggestions. None of these should be implicated in any remaining errors. This book would, undoubtedly, not have been written without the support of my partner Anne Peake.

Part I

Introduction to Building Economics

Chapter 1

A Macro and Micro Economic Introduction

Introduction

This chapter attempts to survey the field covered by 'Building Economics' and to address some areas previously neglected in the literature of the subject. We will begin by asking what building economics is and whether it is, as some assume, a branch of 'general' economics? That controversial question (to those in the field) will lead us to consider, in outline form, the history of economic thought. For our purposes here we will be giving particular attention to the diversity of approach and the concentration on specific issues, which are inherent to many schools of economic thought. This leads us to an examination of contemporary approaches and finally to a brief review of some of the fundamental concepts through which orthodox economists view the world.

What is Building Economics?

Conventional definitions of Building Economics tend to imply that economics is a body of knowledge and theories about which there is general agreement. According to Hillebrandt (1985: 3):

> 'Construction economics is a branch of general economics. It consists of the application of the techniques and expertise of economics to the study of the construction firm, the construction process and the construction industry.'

Hillebrandt is, as usual, concise. Her definition prompts two observations. Firstly, that quantity surveyors and building cost consultants may have a vested interest in elevating what they do to the level of 'economics' with its implications of professional, as opposed to technical skill, and its association with scientific methods, theories, conjectures and refutations. It could be argued, for example, that this claim is not wholly convincing and that building economics has much more in common with, the so-called, cost and management accounting.

To support this contention, we may call as evidence the fact that the objectives of practising economists and building economists are different. Put simply, mainstream orthodox economics is the study of how people and society choose to employ scarce resources, which may have alternative uses, to produce and distribute various goods, services, and factor-incomes therefrom among the members and groups in society. Building economics is conventionally said (by UK based practitioners) to be about helping clients to achieve, frequently mentioned but rarely defined, 'value for money' from their new or rehabilitated buildings. This is sometimes misunderstood to be about cost minimisation. In fact, in both public and private sectors it may be said to be about maximising the difference between the cost of the building to the owner (the price charged by the contractor) and its value, either in use or exchange.

It could be argued that these two objectives are not compatible. Further evidence to support this cost accounting contention may be found by consulting key (UK) texts on building economics such as Seeley (1983) or Ferry and Brandon (1991). The arguments do not stand up however, building economics, at least as it is practised in the UK, is a branch, if of anything, of microeconomics. It is concerned with identifying optimal allocations of resources for building owners or developers. This is undoubtedly much more than mere cost accounting.

Decisions concerning the appraisal of and allocation of resources to buildings and civil engineering projects take place in a relatively long time frame. The time taken to conceive of, and then design and construct such a facility is usually measured in years. The benefits from the project do not normally arise until it is in operation. Thus, in order to initially appraise the project, the eventual input costs and benefit stream will have been forecast a long time in advance. Such forecasts need to take account of dynamic economic conditions and technological advances. They cannot be made by mechanical application of formulae. The building economist needs to be able to adjust forecasts to take account of changes in underlying conditions.

Optimal allocation of resources for an individual *vis à vis* one project will not necessarily lead to an overall optimum across the portfolio of investments for that individual, nor will such a local optimum necessarily have any direct bearing on the optimal allocation for society. To contend that building economics is primarily about a combination of technical skills, informal optimisations, cost accounting, cost control, price forecasting and resource allocation, is not to diminish it in any way, for these are important and necessary if we wish to advance our individual and collective economic welfare. Nor is it to suggest that it is possible to practice these skills in the absence of any knowledge of economics. On the contrary, it is crucial that any practitioner of building economics has a thorough knowledge of economising behaviour, of the workings of the economy and, most importantly, of the consequences of economic policy decisions.

The second observation, prompted by the definition is that it does not (nor does it have to) explicitly acknowledge that there is much disagreement amongst economists on techniques, theories, methods and models of real world activity. The encroachment of mathematics into twentieth century economics gives the illusion of precision and has led one distinguished applied economist to conclude:

> 'The economic journals indicate that many of the most prestigious academic economists are working on theoretical mathematical models that begin with entirely arbitrary and unrealistic assumptions and lead to precisely stated and irrelevant conclusions as far as the real world is concerned.'
>
> (Kamarck 1983: 122)

Economics is a social science and not a 'pure' science. It is not usually possible to perform laboratory experiments. There are many untested and untestable models. There are divisions between different schools of thought.

Building Economics and Qualitative Reasoning

However, we must be careful to avoid rejecting the positive attributes of economics. The real world is sufficiently complex that without some conceptual framework, some tests of logic, some theoretical underpinnings, a thinking individual will drown in a morass of data. Ideally of course, it would be convenient to have precise quantitative reasoning for every decision. Here however, we take the view that qualitative reasoning is frequently sufficient to form a basis for informed decision making. For example, it will be important for construction professionals to know whether some variable (interest rates, public spending, savings ratios, real earnings, inflation, for example) is likely to increase or decrease; what the consequences of this are and whether they are likely to be large or small.

For informed judgements we need often to be aware only of the direction and general order of magnitude of changes. In order to achieve this, we need to be able to establish causality among economic variables and their relationships with government decisions. This approach necessitates a familiarity with both the built environment and its process of production and also with the strengths and limitations of economics as it is currently understood. On this latter point we commence by briefly reviewing the history of the development of economic thought.

The History of Economic Thought

Overview: Economics, Value Judgements and Ideologies

John Maynard Keynes was convinced of the immense influence of ideas as is shown by this famous quote:

> '... the ideas of economists and political philosophers, both when they are right and when they are wrong, are more powerful than is commonly understood. Indeed the world is ruled by little else. Practical men, who believe themselves to be quite exempt from any intellectual influences, are usually the slaves of some defunct economist. Madmen in authority, who hear voices in the air, are distilling their frenzy from some academic scribbler of a few years back.'
>
> (Keynes 1936: 383–4)

The development of economic thought took place (and continues to take place) in a discontinuous way. The questions examined by the various schools are usually those which are of contemporary relevance at a particular point in time in a given set of circumstances. The analysis of one question and thus the exclusion of some other, is therefore a form of value judgement. Karl Marx and Friedrich Engels pointed out the existence of 'ideological bias' in their criticisms of classical economics. By this they meant that peoples' ideas are likely to glorify the interests of those groups (or classes) in society that are in a position to exert themselves and thus to lead to analyses and conclusions which diverge from the 'truth'. They themselves appear to have been exceptions to this rule.[1]

In the medieval universities there was a finite amount of knowledge, of the world, the arts and sciences, which could be taught. As time went on, it became clear that as new discoveries were made and older theories refuted, the sum of knowledge changed, mainly by being increased. New paradigms evolved which reflected the consensus view. We moved from notions of a flat earth around which everything else revolved, to the idea of a spherical earth revolving around the sun.

At the beginning of this century we moved away from Newtonian deterministic physics to a paradigm based on relativity and the notion of uncertainty and probabilistic, as opposed to deterministic, descriptions of the world. One objective of the review of the history of economic thought outlined below, is to indicate the main developments in our knowledge of economics. Additionally, and more importantly, we will demonstrate that the position we occupy now, is conditioned by our history and by social and economic conditions in the late twentieth century, and as such is a transient position.

Problems of public administration, agriculture, commerce and finance were discussed and written about in ancient China and, more recently, in Graeco–Roman times. According to Schumpeter (1954: 51–2) the most important writers were Confucius (551–478 BC), Meng Tzu (Mencius, 372–288 BC), Plato (427–347 BC) and Aristotle (384–322 BC). Detailed histories of economic thought sometimes start from this position. For our purposes here we will begin with the time of the industrial revolution and look at five distinct 'ages', Classical, Marxian, Neo-classical, Keynesian and Post Keynesian and finally, some contemporary alternatives.

Classical Economics

The four major contributors to classical economics may be usefully placed in the context of the Industrial Revolution. In relation to which, Adam Smith wrote before, David Ricardo and Thomas Malthus during, and John Stuart Mill after (Dasgupta, 1985: 11–19). Before considering each of these authors in their turn, let us consider the characteristics of Classical Political Economy.

Four main features may be identified as follows. Firstly, a dominant concern with the primacy of investment or 'capital accumulation' as it was known then. Secondly, the classical economists dealt with aggregates, what we would call today a macro-economic level. Thirdly, the concept of 'value', and a measure for it, was important for their approach, as it was this measure which would enable them to convert the corn, coal, cattle, cotton and the various goods bought and sold, into some form of common unit. In other words, to deal at an homogenous aggregate level with all of the various and heterogenous goods in the economy. Fourthly, they all took as given, the existence of classes in society, primarily the property owners and the labouring classes.[2]

The fundamental problem of the classical economists is best summed up by the title of the book where the 'invisible hand' made its first appearance. In 1776 Adam Smith published *An Enquiry into the Nature and Causes of the Wealth of Nations.* Smith, who had been professor of Logic and then of Moral Philosophy at Glasgow University, was well read in both the social sciences and humanities. It has been said that he was more of a synthesizer than an original thinker (Landreth 1976: 36; Schumpeter 1954: 182–193). Be that as it may, he is popularly held to have been the principal originator of the school known as 'classical economics'.

The *Wealth of Nations* was a 900 page work which combined historical description with deductive theorising. This work was the first comprehensive description of the interrelationships among the elements of an economy. Smith, in keeping with the prevailing world-view, believed in a 'natural' order. He is remembered mostly for his analysis which showed that the unfettered operation of free markets leads to a minimisation of profits and to an optimum use of resources. This, quite naturally, made him popular with the business interests, the capitalists. Beginning with the assumption that humans are rational, self-interested beings, Smith asserts that if left alone a man or woman will pursue his or her self-interest and that in so doing he or she will promote the good of society. Less well known however, is the fact that Smith actually advocated Government intervention in certain cases. These included: the protection of infant industries, trade regulation where matters of national defence were involved, and the provision of roads, schools, statistical records and justice.

Smith was primarily interested in policies which would foster growth. He held that the labouring classes could not accumulate capital because the level of wages was so low as only to allow subsistence. The landowning classes had the potential to accumulate capital, but spent it instead on excessive and unproductive consumption in their rich lifestyles. The rising industrial classes, however, held the key to growth as they accumulated capital and then used it for saving and investment. Dislike of the landowners was to be a recurring theme among the classical economists. His analysis led to two major policy conclusions. Firstly, that governments should operate a policy of *laissez-faire* [3], and secondly, that the distribution of income should be unequal in favour of the capitalists. These would lead to an optimum use of resources and to continued economic growth as the accumulated capital was reinvested.

It is worth noting here that Smith's advocacy of *laissez-faire* was conditional on the existence of free and competitive markets. This has contemporary implications for the privatisation of many semi-state operations in the UK in the 1980s. Put simply, it is not a good idea to deregulate a monopoly in the absence of some approximation to perfect competition. It seems unfair to leave Smith at this point and especially without describing his work on productivity and the division of labour for which he used the famous example of pin manufacturing. We will however explore his treatment of the concept of 'value' in some detail in the next chapter.

Twenty-two years after Smith's *Wealth of Nations*, Malthus published the first edition of his essay on the *Principle of Population*. This was to run to seven editions published at intervals up to 1826. Malthus' thesis went broadly along the following lines. Given, that food is necessary for survival, that sexual passion is necessary and is not likely to change in intensity, and that there is a relatively fixed amount of land available for agricultural production, then the population will tend to grow at a faster rate than the food supply. According to Malthus, food supply would increase in an arithmetic progression while population would increase in a geometric progression. His contention that food supplies could not keep up with population growth caused considerable controversy at the time. Concern with the 'population problem' was not a random event.

After 1790, Britain had ceased to be self-sufficient in food production, food imports had begun and food prices were rising. Secondly, the continued growth of factory, as opposed to home, production and the accompanying movement of the population into concentrated urban areas had led to increased misery amongst poor, displaced people. Some historians argue that there was no decrease in the standard of living but that urban poverty is more noticeable than rural poverty. In either case, population and poverty was perceived to be a contemporary problem so it is not surprising that theories of population began to appear and to be discussed.

Malthus' population theory was important for two reasons. Firstly, its gloomy predictions focused minds on the problems of poverty, wages and population. Secondly, as the essay went into later editions the methodology became sharper, with statistical data and a more scientific approach. The flaws in the theory are, of course, easy to identify two hundred years later. In particular, the fact that Malthus did not take any account of technological innovation in increasing agricultural output. More importantly, Malthus failed to realise that it would be increasing (not falling) income which would slow down population growth. This was not obvious at the time and the central conclusion to be drawn from the Reverend Malthus' propositions was, that over time, labour wages would move inexorably towards subsistence levels.

Ricardo (1772–1823) made a personal fortune as a stockbroker and a fundamental contribution to the history of economics. The Napoleonic wars, a series of bad harvests and rapid population growth had lead to Britain becoming a net importer of grain and to the introduction of the corn laws. The corn laws and the absolute protection they provided for British agriculture were matters of great controversy. Barber (1967: 76) refers to a tenfold increase of landlord's return on one estate between 1726 and 1816.

Ricardo has been called the theorists' theorist. He made significant contributions to the methodology of economics, the theories of value, rent and international trade, public finance, and the concept of diminishing returns. Some people say that his undoubted fame and the esteem in which he was held by his followers was due to the fact that he was telling them things they wanted to hear. He is said, by Schumpeter (1954: 473), to have benefited from the felicitous combination of adopting popular policies which gain supporters, who then defend the methods used to arrive at the policies.[4] His contribution to methods consisted mainly in taking a more theoretical abstract approach than Smith. His prose style was terse. He identified simplifying assumptions and, given these, he reached particular conclusions by implicit reasoning. His approach was more scientific (although still prone to error) than previous writers. Ricardo's major work *Principles of Political Economy and Taxation* was published in 1817.

Witnessing the rapidly escalating income of the landowners, Ricardo shifted the focus of classical economics. Where Smith had felt that the purpose of the so called political economy was to establish the causes of growth, Ricardo set out to uncover the factors which determined the distribution of wealth among the various classes in the economy. Ricardo, like the others in this period, held that there were three main groups in society, the labourers, the capitalists and the landowners. He used the notion of a 'wages fund', a fixed amount of money which was available in the economy for the paying of wages, and Malthusian population theory to show that, in the long run, wages would tend towards bare subsistence level. His other controversial conclusion was that over time profits would decrease, rents would rise and that there would be a decrease in the rate of economic growth. No wonder that about this time economics became known as the dismal science.

Ricardo's contribution to the economic analysis of 'land rent' or 'economic rent' is worth pursuing here for a reason which will become obvious. The term 'economic rent' is most unfortunate and requires clarification. It is important to distinguish between 'rent' and, what we shall call, 'rental'. The latter is a payment made for the (temporary) use of some good or piece of equipment, for example, an apartment, a hire car, a tower crane. Economic rent, on the other hand, is a payment made to a factor over and above that which is necessary to prevent it from transferring to another use. Consider, as an example, a government planning department which increases the salaries paid to civil engineers in order to attract more engineers from industry into government service. Those engineers who move into the planning department will receive only transfer earnings. Those engineers who were already content to work in the department will also receive a salary increase which, to them, will be economic rent. Unfortunately, many economists drop the adjective *economic* thus causing confusion with the concept of *rental*.

The high price of corn in early nineteenth century Britain was the subject of great controversy. Many people held the view that the price of corn was high because the landlords were charging high rents for their land. Therefore the farmers had to increase the price of corn in order to pay the rents. Ricardo, on the other hand, elaborating on one of Smith's ideas, held that the price of corn was high because it was in short supply. The short supply and high price of corn encouraged competition among farmers, who then bid up the price of land in their efforts to reap profits from growing highly priced corn. The price that landlords could charge for their land depended upon demand and supply, and supply was fixed. Therefore, if the price of corn fell, the demand for corn land would fall and so would its price. Making the assumption that land has no alternative uses, it followed that it was not necessary to pay anything to prevent it from transferring to other uses. Therefore the rent paid for the land was a surplus over and above the transfer earnings (zero in this case). Hence, the use of the term *rent* to describe this surplus. Thus, Ricardo concluded that land rent was *price determined* and not *price determining*. It follows therefore, that the demand for land is *derived demand*. This is the basis for what we now call the *residual method* of land valuation, an important tool in the development appraisal of any construction project.

Mill (1806–1873), the last of the classical economists which we shall consider here, differed significantly from the others. Writing at the end of the classical period, commentators have found him difficult to classify. Some say Mill represented the decadence and decline of classicism, others, that his was the most mature statement of classical economics. Some things are certain, he was an eclectic thinker who was at least as concerned with the social order as the economic order and that he made significant revisions to classical economics. In Mill's time there was a growing divergence between the conclusions of economic theory and the actuality of economic fact. In particular, Ricardo had predicted that returns would decrease over time, in fact agricultural production was increasing, and there was evidence that wages were increasing.

The Ricardians were at this time becoming increasingly associated with strongly conservative views. The doctrine of the wages fund was being used to support the position that it was impossible to attempt to increase the wages of labourers to a level above subsistence level. Mill recognised more clearly than anyone before him that the workings of free markets did not necessarily lead to a 'natural order' economically or socially.

Mill differentiated between two types of 'law', on the one hand, the immutable natural laws and on the other, laws which governments or institutions could influence. The laws of production Mill held to be of the former type, and the laws governing distribution of income, of the latter. According to Mill then, it was not at all inevitable that the wages of the workers would remain at subsistence level. Towards the end of his career he withdrew his support for the doctrine of the wages fund. This removed an important prop from those who believed that the masses were condemned to a permanent subsistence living and that the formation of labour unions was pointless and unnecessary. Mill advocated the formation of producer cooperatives and what might today be called market socialism. Mill's *Principles of Political Economy with some of their Applications to Social Philosophy* was written over 18 months and published in 1846.

Mill left a lasting imprint on the earlier work of Smith and Ricardo and he went on to dominate economic thinking until the end of the nineteenth century.

Marxian Economics

Many historians of economics agree that Marx (1818–83) was a classical economist. According to Schumpeter (1954: 383), '... Marx was part-and-parcel of that period's [1790–1860, the period of Ricardo, Malthus and Mill] general economics'. That he reached very different conclusions to the classical economists is well known. Less well known is the fact that in order to do this he made use of much, if not all, of the analytical framework laid down by Ricardo in particular. The ideological content and commitment of Marx and of his followers has the consequence that it is very difficult to deal objectively with his economics. His importance in the history of economics is one of the things agreed upon by mainstream orthodox economists.[5] We shall deal with him separately here for the central reason that his perspective was sufficiently different and has given rise to a world view held by his contemporary followers which is quite a long way removed from mainstream orthodox ideas, that he warrants separate treatment for our purposes.

Marx's university education was as a philosopher. On the evidence of his writings he has the equal right to be regarded as a historian, sociologist and economist. His knowledge of literature is said to have been exhaustive. Ironically, in view of his proposition that each class tenaciously pursues its own interests and his own espousal of the proletarian cause, he came from a distinctly bourgeois family. There is a further irony, in view of the preoccupations of contemporary Marxians, in the fact that *Das Kapital* was about capitalism not socialism. Marx's objective in this, his major work, was to identify the cause of motion of the capitalist mode of production. This objective gives a clue to his general approach, which was teleological, that is to say, he believed that events moved inexorably to a pre-ordained conclusion. This is one feature which set him apart from previous writers with the exception of Hegel. Smith, Ricardo and Mill had all written with the intention of reaching policies which if enacted would lead to a betterment in economic conditions. Marx, on the other hand, believed that human intervention in the inexorable forces of history was ultimately futile.

In view of the panoramic nature of Marx's writing it is probably unfair to extract only the economics, but given our purpose here, there is little choice.[6] Marx's economic theory is an application of his theory of history to the capitalist economy. One of his most important central propositions is that there are fundamental conflicts of interest between the classes in society. These conflicts lead to antagonistic social relations between the classes. Marx believed that he had uncovered immutable laws concerning the development of history through stages. He believed that economic laws were only true for their own particular stage in history. He began with a labour theory of value which owed much to Ricardo. From this he developed Marxian economic 'laws' and some prophecies, amongst which were the following: the existence of a 'reserve army' of the unemployed which exerted downward pressure on wages and so ensured that profits never reached zero. Marx, like Smith and Ricardo before him, believed that profits would decrease over time. He believed that periodic fluctuations and instabilities would lead to business crises. He also believed that there would be an increasing concentration of industry into fewer firms and that the proletariat would become increasingly miserable as their jobs became more dull and repetitive. *Das Kapital*, Volume I was published in 1867.

In conclusion, the concept of 'social relations' (people interacting in their social and economic context) is central to the Marxian view. Although this may all seem rather remote from the economics of building design, appraisal and production, an interesting, useful and thought provoking recent Marxian contribution to the literature on the economics of the construction industry is Ball (1988).

Neo-classical Economics

The Victorian age was, in one of the British Isles at least, an age of relative prosperity. The gloomier predictions of the classical and Marxian writers had not come to pass. The intellectual current was one of faith in the benevolent consequences of progress. With hindsight, we can see that the time was ripe for change in the prevailing climate of economic thought. The neo-classical age, as it became known, was characterised by five distinctive features.

Firstly, political economy became more international. The man usually associated with the elucidation of neo-classical principles, Alfred Marshall (1842–1924) was living in Britain but important contributions came from, Leon Walras (1834–1910) in Switzerland, John Bates Clark (1847–1938) an American, Knut Wicksell (1851–1926) in Sweden, Carl Menger (1840–1921) in Germany and the Austrian, Bohm-Bawerk (1851–1914). A second superficial characteristic is that there was a general shift of concern away from the big 'Macro' questions such as the causes of economic growth, the pattern of long-period change and the ultimate end point, the telos, of the economic system.

The remaining characteristics are more detailed and more important. Thirdly, neo-classical economics saw the introduction of marginal analysis, a fundamental element of modern orthodox economics, this will be described below. Fourthly, the specific focus of neo-classical economics was on the processes by which markets allocated resources. This led to conclusions much more optimistic than those of the classical writers. Although both groups, using widely different types of analysis, held the view that unregulated markets did not necessarily lead to the optimum use of resources. Fifthly, the work of Walras in particular was highly abstract and made use of mathematics on an unprecedented scale.

This last point had good and bad results. On the plus side it certainly tightened up the arguments and it became less easy for the illogical and the value-laden to masquerade as the scientific. On the minus side, increasing rigour demanded explicit assumption, the increasing complexity of which resulted in conclusions which were difficult to relate to the rather untidy, 'real world'. Walras's intention was to produce an exhaustive account of the complete implications of perfect competition. As 'perfect' competition is but a concept which although analytically useful, rarely occurs in practice, he was understandably criticised. Barber

(1967: 2) quotes Walras in defence, 'what Physicist would deliberately pick cloudy weather for astronomical observations instead of taking advantage of a cloudless night' (Walras, 1954: 86). That said, Walras may have left open the way for some valid criticisms of unreality and intractability such as that voiced by Kamarck (1983: 122) above.

In the neo-classical view of the world, the point of the economic system was not to produce commodities, it was to produce 'satisfactions', or utilities. This had two important consequences. Firstly, it lead to a much more realistic view of the economy as a whole. There was no longer a distinction between commodities. It was now possible to include services in the workings of the economy. There was no need to treat disdainfully the shopkeeper who sold the furniture, he or she has as valid a role in the system as the cabinet maker who makes the piece of furniture. Each produces utilities.

Secondly, the concept was developed to include the notion of diminishing marginal utility. This relates to the utility derived from each additional unit of consumption. Consumers gain utility from goods but the amount of utility gained reduces as they buy more of the same goods. A chocolate-loving-consumer gains less utility from the fifth bar of chocolate than from the first and is consequently willing to pay less for the fifth than he or she might be willing to pay for the first. Marginal analysis was at the core of neo-classical economics and has subsequently been refined and developed and is central to mainstream economics today. There are lots of 'marginals', revenue, cost, product, utility and so on. Marshall used this basic idea to develop a wide range of theory. Here is an example from the theory of the firm. According to Marshall a profit maximising firm will continue to expand output up to the point where marginal revenue (MR) equals marginal cost (MC). The logic for this is that if the MR gained by selling an extra widget exceeds the MC of producing it then it is worth expanding production. On the other hand if MR is less than MC then the firm will lose money by producing the extra unit.

Consumers can maximise their utility from a given income by varying the combination of goods and services they buy. Rearrangement of spending patterns will be achieved by substitution of goods. Therefore the demand for goods which have close substitutes will be sensitive to price. The notion of land, labour, capital and entrepreneurship as factors of production gained currency in this time. It was also assumed that factors of production would have alternative uses, leading to the concept of 'opportunity cost'. This has contemporary application, for example, if a firm decides to finance a new investment project from its own profits the finance is not 'free'. Its real cost is the amount that the firm has had to forego by not using the money for some other purpose, for example, putting it in the bank and earning interest.

Finally, it was in the neo-classical era that the prime importance of the concepts of supply and demand in determining prices was first worked out. Market prices would tend towards an equilibrium point. A price below equilibrium would lead to rapid clearing of the market. Some frustrated buyers would bid up the market price towards equilibrium. Conversely, a price above equilibrium would result in sellers offering more units than buyers would be willing to take. Sellers would then compete with each other, reducing prices to the level at which the market would be cleared. Marshall used the analogy of a pair of scissor blades to demonstrate that it was not possible to say whether value (which he took to be synonymous with market price) was governed by utility or by cost of production.

Before leaving the neo-classical era we must refer to the work of the so-called 'Austrian School' and of Carl Menger and Eugen Von Bohm Bawerk in particular. Although Bohm Bawerk's work focused on the nature of capital and interest, the range of his theory was so wide that no less an authority than Schumpeter (1954: 846) has called him the 'bourgeois Marx'. From the perspective of building economics we are interested in the premises of Austrian subjective value theory in so far as they have the especially useful characteristic of encompassing a time dimension which has specific application in building life-cycle benefit and cost appraisal. In this perspective it is assumed that maximisation of utility is the key criterion for economic decision making. Present and future satisfactions need to be weighed against each other. However, a bird in the hand is, as they say, worth two in the bush. Therefore we tend to overvalue the present compared to the future, and to underestimate future costs.

This implies that we need to be rewarded, via a rate of interest, for foregoing present enjoyment in favour of potential future satisfaction. From this point on, a rate of interest is a valid payment/reward without any of its classical associations with 'usury'. Austrian theory is currently being employed and developed to provide a theoretical foundation for building economics (Bon 1989).

Keynesian and Post-Keynesian Economics

In the 1920s and 30s the reality once again began to diverge significantly from the predictions of conventional theory. Whereas neo-classical thinking had deduced that economic equilibrium would be reached at full employment, unemployment had risen to unprecedented levels. The economies of Britain and the USA were in crisis, bread lines and soup kitchens had to be provided to feed the hungry. Some people began to wonder whether the Marxian prognosis about the ultimate end of capitalism was about to be borne-out after all. Neo-classical orthodoxy was certainly not equipped to deal with this situation. It was during the 'Great Depression' that John Maynard Keynes (1883–1946) formulated his 'General Theory of Employment, Interest and Money'. Keynes had first risen to international public attention when in 1919 he had published *The Economic Consequences of the Peace,* a sharp attack on the retribution payments exacted by the allies from Germany.[7] It was said of him afterwards that Keynes had spoken '. . . when men of equal insight but less courage and men of equal courage but less insight kept silent' (Schumpeter 1954: 1170).

The central question of Keynes' work was a direct challenge to neo-classicism. Keynes demonstrated theoretically that, not only was economic equilibrium at full employment not guaranteed, but that in the possible range of outcomes it was highly unlikely. That said, he felt that the capitalist system was probably the best route to the good life but that it contained features which were objectionable and that it would be preserved only if there were appropriate government intervention to prevent social unrest generated by mass unemployment. Unsurprisingly, he was the subject of criticism from both right and left. From the former, for advocating government intervention and from the latter for his defence of the market system.

Keynes' work was at the macro, aggregate level of the economy and he was particularly interested in short-run changes. He is said (by Pigou) to have observed that 'in the long-run we are all dead'. Keynes played down the importance of interest rates and supported fiscal management, i.e. manipulation of tax rates and public spending. In particular, he advocated deficit budgeting, the benefits of which he was able to demonstrate by using the concept of a 'multiplier'.[8] This indicates that an increase in investment generates increased demand and income as workers are paid and spend their incomes creating waves of expenditure and income radiating out through the economy.

Keynesian theory laid out the principles governing the managed welfare-oriented capitalism which existed in most western industrial countries during 'the long boom', 1951–73. The late 1970s and 1980s saw moves away from government intervention which will be explored below.

Contemporary Alternatives

In terms of economic welfare the developed world at least, is undeniably a better place than it was two hundred years ago. It is also true that economists have not yet found a way of running an economy at full employment without inflation or damaging business cycles. There is much to be done. The opinion leaders of the four economic schools described above, were each in their own way, revolutionaries. They promulgated opinions which opposed contemporary received wisdoms. It is in this light that we should view the contemporary approaches within and outside the mainstream. In this section we will briefly consider mainstream orthodox economics as it developed in the 1980s. This will include monetarism and post-Keynesian demand management and the fringes of supply-side economics, rational expectations and Chicago Libertarianism. We will conclude by considering radical approaches and the so-called, Other Economic School of Thought (TOEST).

In the second half of the 1970s and the early 1980s the school of thought known as monetarism occupied the limelight in the economic policies of both the UK and the US. Put simply, this held that if monetary controls were exercised such that money supply grew at the 'right' rate it would be possible to control both inflation and have full employment. Inflation could be reduced by slowing down the rate of growth of the money supply.[9]

Targets for money supply were set by both governments. In the UK in particular, the targets were spectacularly exceeded, monetary growth being fuelled by, among other things, increased public expenditure. Although never really admitted to in public, the policy was dropped in both countries.[10] This led to the rise of 'supply-side economics'. The so-called Reaganomics was a combination of monetarism and supply side economics.[11] As the name implies supply-side economics revolves around the importance of freeing up the productive aspects of the economy, mainly by deregulation or taking the government 'off the backs of the people and out of the economy'. This concept was extended to include the notion that if tax rates were cut, tax revenues would somehow increase. The supply-side view is essentially optimistic, if we get governments out of the way, the freer markets will lead towards equilibrium resulting in an efficient use of resources. Contrast this with the rational expectations school which holds that only surprises matter. Rational decision makers take into account present and expected future conditions. Government intervention is therefore of no avail, as people can build these new conditions into their assumptions and work their way around them (Minford and Peel 1983). Related to this, is the so called Libertarian school which deduces that personal freedom is enhanced by minimalist government intervention and the free workings of markets (Hayek 1960).

These approaches are in direct conflict with the ideas of Keynesian demand management, into which the USA government retreated in 1982. This involved 'spending your way out of recession', deficit budgeting to increase demand disproportionately via the multiplier, and to stimulate employment on the supply side. The remaining problem is that Keynesian economics was designed for emerging from depression but deficit budgeting cannot be sustained indefinitely (Stewart 1972).

While the radical economics discussed below has not been put into practice at a national level, some orthodox economists say that this is true to a lesser extent, of mainstream economics. David Henderson, head of the economics department at the Organisation for Economic Cooperation and Development (OECD), coined the term 'Do it yourself economics' (DIYE) in his 1985 Reith lectures (Henderson 1985). DIYE refers to the intuitive nature of many economic policy assumptions and decisions made by governments. In DIYE for example, manufacturing industry is in some way superior to services, whereas we have seen above that in terms of orthodox theory they both produce utility and added value and there is no reason, in theory, to accord a higher status to either.

Perhaps with the exception of the Keynesians, the other orthodox schools all tend to the view that the welfare of society, 'the common good', may be enhanced by increasing the potential for personal gains through for example, deregulation, privatisation and tax reductions. This is in contrast with the radical schools, one of the most articulate representatives of which is Lester C. Thurow. Although it is not the intention here, many would agree that Thatcherism was radical.

Thurow condemns the very foundations of what he terms the

'equilibrium price-auction' (orthodox) model of the economy. The basis of his criticism is that the equilibrium price auction model rests on the assumption that the economic person is a rational utility maximising individual. This behavioral assumption was abandoned by sociology and psychology in the nineteenth century. Rational, valid and useful economic decisions are governed as much by habit, duty, power, self-esteem, altruism, trust and security as they are by utility-maximisation.[12] A radical economist would say that the equilibrium priceauction model, apart from being a static and deterministic model of dynamic and uncertain world, is generally an over simple income and utility-maximising model of a complex world of human interrelationships.

A more diverse and more radical critique of orthodoxy has emerged in the 1980s known as 'the new economics'. The new economics has been developed by a number of organisations and individuals but it is most closely associated with the work of The Other Economic Summit (TOES), (TOES 1984, 1986). New economics attacks the necessity for the pursuit of conventionally defined economic growth, which it says in orthodox economics has become an end rather than a means to other objectives.[13] There is not as yet a fully worked out theory of new economics but the proposals which are emerging are based on the satisfaction of basic human needs rather than on utility maximisation.

New economics tends to be biocentric (concerned with the continued welfare of the planet and its resources) rather than anthropocentric (concerned with human beings). New economics takes up the criticism of Thurow that economics is too narrowly defined and incorporates sociological and psychological perspectives. For example, the human needs which form the core of new economics have been classified by Max-Neef (1986) and separated from the satisfiers of needs. Max-Neef suggests the following nine fundamental needs; permanence (subsistence), protection, affection, understanding, participation, leisure, creation, identity (or meaning) and freedom (Max-Neef 1986). Thus, housing, food and income are satisfiers of the need for subsistence. Defence, cure and prevention are satisfiers of the need for protection and so on. According to Max-Neef, fundamental human needs are finite, few and classifiable, and furthermore, they are the same in all societies. What differentiates cultures from each other and over time, is the means used to satisfy the needs. Thus, one may belong to a consumer or an ascetic society. One measure of cultural change is the progression to new satisfiers of the fundamental needs.

Finally, some of the ideas of new economics have gained support from mainstream thinkers, an example of this is the notion of a 'social dividend'. Orthodox economic policies involve little or no payments to child rearers, home makers, home based carers for the sick and the old. Effectively, they are placed extremely low on the list of societal priorities. The social dividend is a basic, unconditional subsistence payment to all citizens. Among other things, this would encourage self-employment and flexibility in employment patterns and has found favour with Britain's economics Nobel Laureate and Professor Charles Handy of the London Business School (Handy 1988, Meade 1975).

Some Fundamental Concepts

Thus, the body of thinking known as 'economics' has developed, primarily over the past two hundred and fifty years, through many differences of opinion and emphasis, through testing and development of previous work and through responses to pressing social problems of the day. The range of contemporary economic concerns from the fight with inflation and unemployment through to green issues, is no exception.

In this light, let us consider, very briefly, some key ideas which are important for the study of building economics.

Markets in Theory

As we have seen, the theoretical construct of the perfectly competitive market began with Adam Smith. The characteristics of a perfect market are: a large number of buyers and sellers, so that each firm produces a very small proportion of total output; identical products on sale; free entry to the market; perfect knowledge of the prices and quantities sold in all transactions. Clearly, this type of market almost never occurs in reality. In reality we have various degrees of imperfect competition, caused by, for example, having one or a small number of sellers, or product differentiation on grounds other than price.

Price Determination

In the context of a market, prices are determined by the interaction of demand and supply. Demand for goods is influenced by the price, by the real income of the purchaser, the size of the market, the price and availability of close substitutes, and by personal preferences. The amount of goods supplied is influenced by the price obtainable relative to the costs of production (potential profit). Thus, supply can be influenced indirectly by any change in market conditions which affects the price obtainable and also by changes in technology which may influence the cost of production.

According to the model, all markets tend towards an equilibrium point where the price level ensures that demand and supply are balanced, i.e. the market is cleared and all goods which have been produced are sold. In reality, of course, very few markets ever reach equilibrium but the model is useful, in that it can be used to deduce the consequences of changed conditions. Most of the time it appears that markets and economies are going into, or coming out of boom or recession, or recovering from some external shock or other, e.g. OPEC price changes, wars, famine, political uncertainties, government regulation (pollution control), deregulation, privatisation. Thus, although the markets rarely arrive at their destination the model is useful in pointing out the direction of travel in response to various external stimuli.

Role of Government in a Mixed Economy

Even the classical *laissez-faire* thinkers of the eighteenth century knew that government would need to intervene in markets and economies to ensure that the pursuit of individual self-interest was not counter-productive to the needs of society as a whole. Arguments arose then and exist to the present day, concerning the extent and nature of such interventions. Few would disagree, however, that markets need to be constrained and taxes collected to ensure national defence, pollution control, provision of public administration, provision of care for the elderly the poor and the sick. It is on this last point that great difficulty arises, for politics and economics are inextricably linked. Economists see their role as analysing the consequences of decisions to do with, as we stated at the outset, allocation and distribution of resources. The economic goals to be set and the choice of means of achieving them, need to be identified by society as a whole through their representatives whether appointed democratically or by some other means.

Throughout the 1980s the economic policies of both the USA and the UK, among others, were concerned with increasing the role of markets in mixed economies. This implied reductions in the role of government, based on the assumption that unhindered markets were likely to be more efficient. However, in the late 1980s and early 1990s there was a growing concern with pollution and global issues such as acid rain and global warming. These issues began to force policymakers to confront the importance of the role of government and regulation in mixed economies. As we will see later, the role of government on the economy has direct implications for the construction industry. Thus, from the perspective of building economics, it is important to have some understanding of areas of economics, which may lead to adjustments in this role.

Notes

[1] Value judgements and 'ideological bias' in economics are discussed in detail by Schumpeter (1954), see for example pp. 35–6.

[2] Dasgupta then proceeds rightly, to include Marx in the classical Political Economy epoch. We however, shall deal with Marx separately for two reasons. Firstly, because his work sheds much light on the workings of capitalist mixed economies and secondly, because there is an emerging Marxian approach to building economics. A good example of the latter is in Ball (1988).

[3] *Laissez-faire*: The principle of non-intervention [by government in the economy].

[4] Schumpeter however was sharply critical of Ricardo and his followers and clearly felt that the man was very overrated (Schumpeter 1954 pp. 470–80).

[5] See for example, Samuelson (1985) pp. 768–70, Schumpeter (1954) pp. 383–92, Landreth (1976) p. 155.

[6] The best exposition of Marx's economics is probably Sweezy (1946).

[7] Keynes had accompanied the British delegation to the

Paris Peace Conference as a Treasury Advisor.

[8] This concept had first been formulated by R. F. Kahn one of Keynes' colleagues at Cambridge.

[9] For more detail see Thurow (1985) pp. 74–7 and especially Friedman (1982) and (1968).

[10] This was described almost as it was happening by Thurow (1983: 124–41), and afterwards by Niskanen (1988) and Bosking (1987).

[11] This is described in more detail by Thurow (1983, 142–72) and Minford and Peel (1983).

[12] This argument is developed in detail in Thurow (1983), see in particular pp. 224–37.

[13] On the so-called New Economics generally see Elkins (1986) and (1988) and Miller (1987).

Chapter 2

Concepts of Value, Cost and Price

Introduction

Before proceeding to examine the factors affecting, and methods of forecasting and controlling, building value, costs and prices, it is necessary to explore the nature of the three concepts and their interrelationships. Our purpose is not to review the history of value theory, nor to attempt to assess such detailed matters as the extent to which Galiani and Bernoulli in the early eighteenth century pre-empted the work of Smith and the neo-classicals, on value and marginal utility.[1] We are concerned with the current state of knowledge concerning the concepts.

The overview presented in the previous chapter will provide historical context sufficient for our purposes. Accordingly, this chapter will be in six sections; the first three considering the concepts separately in turn; the fourth considers the relationship between value and price; the fifth considers methods of expressing value, price and cost, and the final section draws together some conclusions.

Value

Value, in common parlance, is taken to mean the intrinsic worth of a good. Most people would allow that it cannot always be measured in money terms, instances of this are items such as archaeological finds and works of art. A work of art may well have a price attached, but it will be shown later (in the section headed Relationship between Value and Price) that price does not necessarily always have a direct relationship with value.

Scarcity affects value. Normally when scarcity increases, so does value. The usefulness of a good also affects value. Hence, the value of a sheepskin coat to a person on a Mediterranean beach in midsummer will be different from the value of the same item to the same person stranded in the Himalayas in midwinter. The value of a good, therefore, gives an indication both of its scarcity and its utility when compared with others. Theoretically, one would expect people to be willing always to pay more for a more useful good than for one which is less useful. Clearly this is not true, and is illustrated by the well-known 'paradox of value' which compares the value of water (low value—high utility) with that of a diamond (high value—low utility).

This paradox arises when one considers utility and scarcity, but in reality there is much more to value than that. Value is the power to serve people's needs or desires. Some of these needs will be subjective, differing from person to person, depending on the varying importance attached to the good in question. It was Daniel Bernoulli who pointed out in a paper written in 1730 that the economic significance of an additional dollar is inversely proportional to the number of dollars already held by that individual. This is of course today known as marginal utility. Thus, it follows that value is not really an intrinsic, inherent property of things, but is a relationship between the valuator and the thing being valued.

This history of value theories did not begin with the classical economists of Adam Smith's era. Aristotle was not only able to distinguish between value in use and value in exchange, but also the fact that the latter stemmed somehow from the former.[2] Adam Smith developed what was to become known as the labour theory of value. His definition of value was based too much on utility, and thus he was not able to solve the paradox of value, about the diamond he said it has 'scarce any value in use; but a very great quantity of goods may frequently be had in exchange for it'. Smith felt that the real price of everything (its value in exchange), what everything really costs to the person who wants to acquire it, is the toil and trouble of acquiring it. If the price was expressed in money or in other goods by barter, then these were merely representative of the amount of labour that was locked up in the good in question.

J. S. Mill began to answer the paradox by taking a much wider (and nowadays generally accepted) view of value. Mill felt in 1848 that the utility of a good lay not just in its power to 'do' something, but also in its capacity to satisfy a desire or serve a purpose (Mill 1970).

The theory of value is still, in this century, the subject of much debate. (Hicks 1948; Dobb 1973) Modern economists hold the view that value in exchange does not come from total utility but rather the marginal utility of the extra increment demanded. In economic terms this can be stated:

> 'The response of quantity demanded to change in price (i.e. the elasticity of demand) depends on the marginal utility over the relevant range and has no necessary relation to the total utility of the good'.

The exchange value of a good is the amount of some other good for which it can be exchanged. Under the barter system, this meant that any particular good had as many exchange values as there were different goods to exchange with it. This has now become rationalised by using a generally accepted means of exchange, which today is money. ('Money' is not the only solution to a 'generally accepted means of exchange' and criteria for this are discussed later in the section headed Expressions of Price and Cost.)

Value, then can be regarded as a complex entity made up of the above components, scarcity, utility, cost of production, value in use, value in exchange and, most importantly, marginal utility. Further, it is also influenced by conditions of demand and quantity available. If you want something desperately, then its value to you is increased. If, as we argue here, value is not an intrinsic quality but rather is a relationship between the valuator and the thing being valued, then its primary determinant is marginal utility.

Cost

The cost of a commodity, whether it be a simple one like a length of timber or a complex item like a building, is the sum of all the payments (whether money or otherwise) made to the factors of production engaged on the production of that commodity. Factors of production are defined in economics as land, labour, capital, entrepreneurship. Land may be defined as natural resources; labour as human resources; capital as money, machinery, plant, equipment and any man-made components not used up in the production process; and entrepreneurship as the risk-taking process of organising the other factors of production.

Costs can be divided into fixed costs and variable costs. The former being those that remain the same over a large range of output; the latter being those that change with every adjustment of output. The term 'cost of production' has meaning only when it is related to output. For example, the cost of a motor car depends on whether the manufacturer is producing 50, 100 or 500 units per week. The term is also ambiguous since it has several different meanings; the cost of production for a given output may be the total cost, whereas for single unit it is clearly the average cost. If a firm is already producing 100 units a week and decides to increase production to 101, the cost of producing the extra unit may be much less than the average cost. On the other hand, if a re-arrangement of production lines and shifts is necessary then it may be much greater than the average costs. The additional cost, whether it is greater or less than the average, is known as the marginal cost.

Whichever particular type of cost is being considered, the essential meaning remains the same, that is, the sum of all the payments to the various factors of production. This appears deceptively simple, it will be shown below that present day methods of calculating 'costs' are often incomplete.

Price

Price is the amount (of money usually) that someone is willing to pay for some particular good. It can either be determined by the interplay of supply and demand in the market situation, or else be fixed by the entrepreneur or the authorities. There is not always a very strong relationship between price and cost.

Consider a highly simplified example; if a development company constructs an office building and the cost is £1 million (i.e., the sum of all the costs paid for the site, professional fees, contractors, etc.). The price it can now expect to sell the building for (or let the space) does not relate to the cost that it has paid for it, but rather to the conditions of demand and supply and how much profit another firm feels it can make by buying or leasing the building. A potential purchaser may calculate that if he or she were to buy the building and manage it and let the space, it would produce an estimated income over a particular time period and that, for this return he or she may be willing to pay say £2 million initially. This may be less or more than the amount the developer wants, but the buyer's calculations for the price it wishes to pay are not based on the developer's cost but rather on the buyer's own potential for making profit. Thus is the case in the short run. In the long run however, as Ricardo pointed out, 'land rent' is *price determined* and not *price determining*. Therefore, in the long run there is a relationship between the developers costs and the market price for built spaces.

There emerges from the above one other important point and that is that one person's price is another person's cost. In short, cost is the amount paid by the buyer, price is the amount received by the vendor.

Relationship between Value and Price

There is a school of thought which holds the view that it is possible to regard the price for a good as a measure of its value in general, making the concept of value and price interchangeable. Price, while in some way linked to value when it is originally set, becomes largely unrelated to value when the forces of supply and demand begin to operate on it. Price can be defined as the money obtainable from a person or persons willing and able to purchase a good when it is offered for sale by a willing seller.

Thus 'price' is equivalent to value in exchange or market value, at a given time. But market value cannot be taken as equivalent to value per se. As has been indicated above, market value or price relates to the degree of marginal utility accruing to the one who demands. This will set the price he or she is willing to pay. Whereas, the value itself in its true sense must relate to the combination of the total utility of the good and the subjective view of the person who is considering it. The value of a building to the owner, occupiers and the community has been proposed as being a function of three qualities: exchange, this would equate with market value as discussed above; utility, or usefulness for its purpose; and merit, the power to satisfy (Powell 1982).

Expressions of Price and Cost

For convenience, price needs to be expressed in terms of a medium of exchange. Under the barter system a good would have many prices in terms of the different things it could be exchanged for. Cattle, shells and tobacco were variously used as money and eventually, pieces of metal of a specified weight and quality. The function to be fulfilled was merely that of avoiding the awkwardness of barter.

Adam Smith laid down an ideological basis for expressing exchange value, 'A commodity which is itself continually varying in its own value can never be an accurate measure of the value of other commodities.' Money clearly does not fit this model, as it is continually changing its value. Using units of labour as discussed earlier does not fit either. The value of, say, one hour's labour depends on whether the person performing it is young or old, whether he or she is working strenuously or at a slow pace; different perceptions of what is 'hard work' among different races, etc. Labour, as a measure, is of little more or less use than money. In this respect Marxian and Capitalist economics face similar problems. On the face of it though, labour as a measure would appear more meaningful than pieces of paper and silver.

It is interesting to consider whether units of energy could be used as a measure of value. These units would represent a sought after and scarce resource. There are many different sources of energy which are, in the main, interchangeable. If any one becomes scarce, prices begin to rise and impetus is given to industries to use alternatives. In the past two decades in particular, political events and violent clashes between cultures have given rise to many short and medium term adjustments in the price of energy sources. However, the only real shortage would be if all resources were in danger of becoming exhausted and world demand were rapidly reaching that point. The long term movement towards the extinction of oil, coal and natural gas supplies can be looked at as a primer for the development of the technology to harness energy efficiently from the sun, wind and waves. If and when this occurs we will have energy sources which, for all intents and purposes, will be limitless. Until this moment arrives we will have a mixture of energy types, some renewable, some not, and this would not be a good vehicle for the expression of value or any of its associated concepts.

Cost is that which must be given in order to acquire, produce or affect something. It has been referred to earlier as the sum of the payments made to the factors of production. However, a consideration of the methods by which accountants and managers calculate 'costs' in modern systems might lead to a redefinition of cost as being 'that which the entrepreneur cannot avoid paying' (Ward and Dubos 1972). Many modern industries still do not include in the cost of what they produce the so-called intangible features of the disposal into the air of effluents or the loading of the land with solid wastes. Unregulated markets treat the environment had as if it no cost. In the words of *The Economist Newspaper* (1989 August 26: 14)

> 'At present most economic activity takes little account of the costs it imposes on its surroundings. Factories pollute rivers as if the rinsing water flowed past them for free, power stations burn coal without charging customers for the effects of carbon dioxide belched into the atmosphere, loggers destroy forests without a care for the impact on wildlife or the climate. These bills are left for others to pick up – neighbours, citizens of other countries, and future generations'.

Not only are these costs unaccounted for by firms, they are ignored in the national accounts of most countries. National accounts take no account of depletion in the stock of natural capital. Conventional national accounting records no increase in assets for a new oil or gas discovery, in fact they record an *increase* in income as the resources are drawn off and sold. Indeed, the cost of cleaning up after environmental disasters such as oil spills is recorded as *growth,* with no deduction for net welfare lost through pollution. It is clear that in order to calculate realistic cost accounts adjustments need to be made for changes in the stock of natural resource and net welfare losses due to pollution (Pearce *et al.*, 1989) [3].

Conclusions

The money paid by the client to the contractor is the contractor's price for providing the project. However, the same money represents only the cost of the project to the client who must include other costs in the larger scheme of things. The client will hope to gain a price for the building space he or she now possesses either by letting the space or by using it and offsetting the rental costs he or she would have had to make elsewhere. The money paid to the contractor is just one of the costs of the building. Other costs the client has to pay include land costs, professional fees, fitting out and general occupation costs.

The contractor's costs are the payments it incurs while erecting the building, this will include wages and salaries to staff working on the particular job, plant costs, materials, interest charges on any monies paid out but not yet received from the client. The nature of contractors' costs is extremely complex, relatively few contractors have an accurate picture of what their real costs are.

Although prices and costs are ultimately interdependent, they are, in the short term, capable of differential movement. There is sufficient statistical data in most developed countries to empirically confirm this by plotting the changes over time in input cost and tender price indices. For example, in the construction market in the London area in the period 1988/89 it was apparent that the consensus of forecasters was that contractors' input costs were increasing at a rate of 5%–7% per annum, while the tender prices charged by contractors appeared to be increasing at a rate of 12%–20%. The explanation for this situation is that while input costs were increasing approximately in line with inflation there was a rapid increase in demand for contractors. Contracting firms found they had more work available to them than they could handle and so were in a position to bid prices up.

This type of differential movement has significant consequences for estimating the future capital cost of projects with data which may only be one or two years old. The short term in this case could be, say, one to three years. If a building project takes even a liberal two years from inception to tender then, for the purposes of tender price forecasting, the short term is all that matters. If a forecaster is using recent building price data to estimate the tender price, say two years hence, of some building currently on the drawing board then it is necessary to update the data by means of a price index. Movement in input costs may be forecast relatively scientifically if there is information on pay awards and general inflation. Forecasting changes in tender price levels is more a matter of judgement and 'knowledge' (probably not numerate) of the market.

Notes

[1] This area is covered exhaustively in Schumpeter (1954: 300–303).

[2] Schumpeter (1954: 56) supports this by means of quotes from Aristotle's, *Politics* I, 8–11 and *Ethics* V.

[3] A more pessimistic view is taken by Mishan (1990: 34) 'Indeed the resulting "market failure" [from spillover

effects] is far too mild a term to denote the almost hopeless inadequacy of the price system today in directing our economic resources so as to prevent injurious environmental effects on a scale that could be moving towards global disaster'.

Chapter 3

The Construction Industry

Introduction

The construction industry supplies services to three distinct types of client. First, firms in both the public and private sector who need building space as part of their production processes. In this case 'production' covers both manufacturing and, services and other non-manufacturing processes. Second, investors who demand building space as part of their investment portfolio and consequently do not wish to use the space for the production of any particular good or service other than that of supplying building space itself. Third, individuals and families who demand housing as, largely, but not entirely, a consumer good. In general, buildings are investment goods. Housing may be regarded as both a consumer and an investment good. Thus, the demand for building space comes from a diverse set of motivations ranging from the first order primary need for shelter, to higher order needs for the production of other goods and services.[1]

The construction industries of many developed and developing countries share similar characteristics. In this chapter we will first introduce a conceptual framework which locates the construction industry in the context of the economy and the market for property. Second, we will describe the economic characteristics of the industry. Third, we will describe the factors which affect the structure of construction industries internationally. Finally, we will describe some operational features of the industry in the UK.

Construction in Context

The construction industry exists because people and firms need shelter in which to carry on their activities. 'Shelter' nowadays takes the form of built space. Just as for other goods, there is a market for the buying and selling of built space.[2] This market has special characteristics. Buildings take a relatively long time to plan and construct. They are relatively expensive goods. For most people the purchase of a house or apartment is the biggest economic transaction they will ever make, costing many times the annual income of the household. Firms are not usually able to purchase new buildings out of annual profits, they need to borrow money for this purpose. Indeed, buildings became objects of exchange relatively late when compared with many other goods. This is unsurprising as, compared with most other goods, they are virtually impossible to move around, and are connected to land which is itself a resource with special characteristics.[3]

The number of buildings on sale at any one time is a small proportion of the total stock. Therefore, for given transactions, there is usually not a large number of buyers and sellers. In some countries, the actual selling price of particular buildings is not revealed to third parties or the market at large. The information which may be freely available is the asking price, this may or may not coincide with the agreed price. Transactions take time to complete. The market does not always clear very quickly. At any one time there may be property stuck on the market because the minimum price required by the seller is above the maximum price prospective buyers are prepared to pay. The market for property has very important locational features. For a building to be of use to a purchaser it must be the right building in the right location.

Built space is heterogeneous. It is usually difficult to compare characteristics and prices of different properties. The prices that purchasers are willing to pay will be related to their expectations of income and/or utility from the space. There is always a measure of uncertainty surrounding such projections. Even if forecasts are perfect there is still the possibility that external conditions will change in the future thus diminishing the accuracy of the forecast. Although there is an underlying rationale, the market for real property is volatile.

The peculiar characteristics of the property market have led to the growth in influence of a distinct class of mediators between buyers and sellers of space, the property or real estate 'agent'. Demand for built space may be satisfied from the existing stock of buildings or by the purchase of a new or rehabilitated building from the construction industry. The demand for built space will only be translated into construction business if there is not already an appropriate building in the right location.

The demand for most buildings is 'derived demand', that is, it depends on the demand for goods and services which can be produced from the building or for the utility offered by the building. Thus, the construction industry exists at the interface between, on the one hand, the supply of existing buildings, each with its own physical and locational characteristics, and on the other, the general conditions of demand prevailing in the economy. This distinction is important, as construction work is influenced, not only by conventional direct demand for production and consumption, but also by demand originating from 'market making' activity in the property 'industry'. We may consider the property industry as those developers and agents who arrange finance, construct, buy and sell property.

Market making derives from the expectations, held by property agents and developers, of future demand for particular types of built space. Of course, the property industry has a vested interest in the success of new markets. These latter expectations may or may not turn out to be correct. For instance, in the early and mid 1980s in the UK there was much optimism in the, so-called, high-tech industries. The majority of the growth in these industries took place in 'Silicon Valley UK' a crescent shaped area running to the North, West and South of London. Land owners and property developers rushed to convert green field sites into new 'high-tech' industrial parks for these new industries. By 1989 it was becoming clear that the property industry had overestimated future demand for this building type (Building Market Report 1989). As more and more travel to work areas reached 'full employment' (i.e., an unemployment rate of less than 3%) new built space was still coming onto the market from the development pipeline with little possibility of finding personnel to work in any incoming industries or businesses. This glut of property was partly exacerbated by the lack of mobility of labour in the UK due to among other things, the north-south divide and large differentials in housing costs not matched by differences in wages (Balchin 1990).

Thus, any attempt to understand the construction industry must be underpinned by a conceptual framework which locates it with respect to the economy and the property market. Figure 3.1 is a schematic representation of this relationship. The demand for built space is influenced by general economic conditions. (Let us, for the time being, accept this simplification, a fuller discussion of demand follows in chapters five and nine.) This demand may be satisfied either through the purchase or

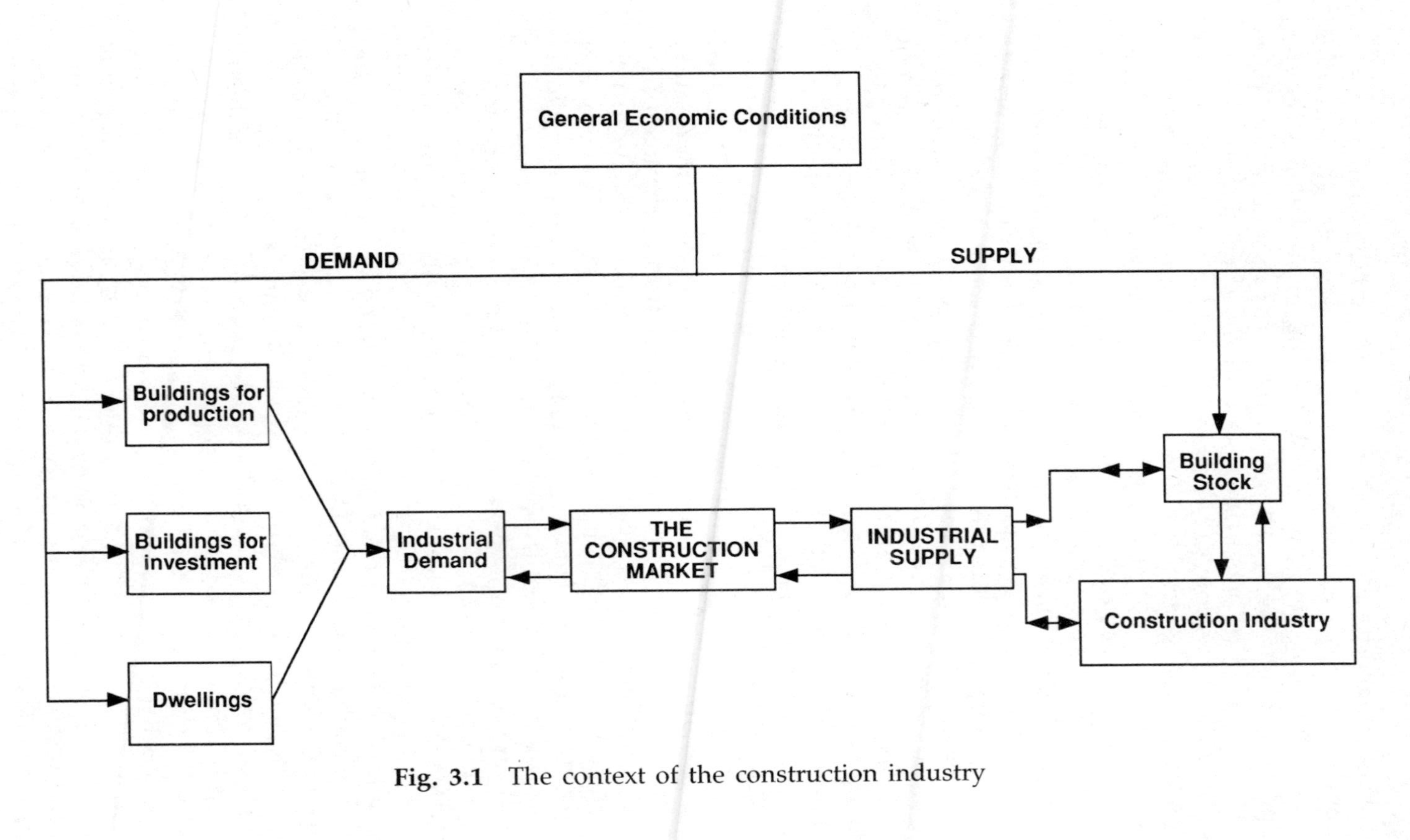

Fig. 3.1 The context of the construction industry

lease of a building from the existing stock or by buying a new or rehabilitated building from the construction industry. Thus, the level of business which is carried on by the construction industry depends on the demand for space and the condition and appropriateness of the existing stock. How appropriate the existing stock is for contemporary use will depend on, among other things, the rate of technological change in industry and business generally. This may have little to do with the physical condition of the stock. For example, even a perfectly preserved nineteenth century industrial building may be of little use for the requirements of industry in the 1990s. Technological, economic and other factors affecting building obsolescence are discussed fully in Chapter six.

Economic Characteristics of the Construction Industry

In the words of Hillebrandt (1984: 1) 'the construction industry has characteristics that separately are shared by other industries but in combination appear in construction alone'. The pattern of demand facing the industry is dealt with separately later. Suffice it to say here that in many developed and developing countries construction output (the nearest proxy for 'demand') fluctuates more widely than GDP as a whole (World Bank 1984: 39). At the same time it should be noted that output of the various building sectors fluctuates more widely than construction output as a whole. The pattern is not universal though.

It is a commonly held belief in the UK for instance, that the construction industry suffers from excessive fluctuations in demand when compared to other industries. Ball (1988: 111) cites two empirical studies which show that this is not the case. The first is a study by Sugden (1980) which calculated the coefficients of variation around output for the years 1950–71. Construction had a coefficient of 2.02, while manufacturing as a whole had a coefficient of 2.04. A study by the National Economic Development Office (1976) showed that construction enjoyed less fluctuation in demand than every other classified industry bar one, food, drink and tobacco. This is a clarification, it is not to underestimate the consequences, particularly for unemployment, of changes in demand for the industry. In the following paragraphs we will leave demand to one side and focus on the economic characteristics of the production of the built environment.

The principal inputs to construction are materials, equipment, personnel, capital and management. It is a feature of production in the construction sector of many countries that there is often more than one combination of factor inputs which will produce the finished project. Alternative combinations usually involve varying mixes of labour and equipment. Similarly, in most countries, construction is influenced primarily by domestic conditions. This is true both in the economic and the technical sense. Technical solutions to the problem of providing built space depend on the conjunction of the climate and the availability of materials locally. For example, well insulated single family dwellings of timber construction (not *timber framed* but *timber throughout*) are suited to the extreme temperatures, but essentially 'dry' conditions, of Scandinavia and Canada, where millions of dwellings of this type were built in the nineteenth and early twentieth century. Furthermore, the timber is available in large quantities at low cost. Even if the material were available at the same economic cost in the UK say, the damp climate rendered this particular technical solution less appropriate. During the same period in the UK the conjunction of climate and the availability of clay resulted in the construction of millions of brick built dwellings.

In addition, building materials are often bulky and heavy and, in general, they have a low value to weight ratio. These two factors tend to ensure that the import penetration of construction materials is generally low. In the UK for instance, the import penetration of building materials is estimated to be 8% or under a quarter of the average level for all manufacturing, (Turner 1987). Although it should be said that individual studies of some large projects in London in 1989 (Broadgate and Canary Wharf) and the advent of the single European Market in 1992 would seem to indicate that the import penetration of building components is rising rapidly. It follows from this that, unlike many other industries, the construction industry of one country is relatively unaffected by the competition from other countries. The fact that this is common knowledge does imply that the situation is unlikely to continue. There is undoubtedly an increasing trend toward globalisation in the construction industry. A further feature of the industry is that it is, relative to others, labour intensive. This felicitous combination of low import penetration and high labour content renders the industry an attractive mechanism for economic regulation.

The historical origins of the industry in trade and self-building skills are still evident today insofar as it is common for there to be a proportionately large number of very small construction enterprises. These frequently comprise a sole working proprietor with, possibly, one or two employees and they may take the form of specialist subcontractors. Manual employment in the industry is, in general, project focused. That is to say, it has been traditional to employ workers for a project and to 'let them go' on its completion. Construction workers do not, in general, enjoy either job security or high wage rates when compared with other industries. It is fair to say that, given the actual conditions of work and the pay, in Western Europe at least, it is difficult to understand why anyone would want to work in the industry if they had any choice. The employment conditions of construction workers in the UK in the 1980s have been described in detail in Ball (1988).

However, one positive consequence of this fragmentation in the industry is that as the firms are differentiated by trade specialism, they are in a strong position to cope with the fluctuations in demand for particular building types. Thus, if the demand for industrial buildings falls, the specialists in specific trades may be able to find work in other building sectors where demand is still buoyant. Thus, although there may be periods of unemployment between projects, the structure of the industry helps to minimise periods of unemployment due to fluctuating demand in specific sub-sectors of the industry.

A number of features of the industry combine to maintain low barriers to entry and generally favourable conditions for the start-up of new firms. First, construction firms have relatively low levels of fixed costs. As the product has to be assembled at the location of demand there is no need for the firm to own and operate a large premises. There is, in the developed world at least, a well developed hire industry for plant and equipment, so these items can be rented just for the duration of the operations for which they are needed. Additionally, as we have seen, there is a well developed tradition of subcontracting and self-employment which results in firms having a relatively low cohort of permanent labour. Second, construction firms have relatively low requirements for working capital. It is usual for firms to be paid monthly 'on account', less a small retention sum, for work in progress. Hence, working capital is needed, at most, to finance one month's work on a given project.

Another factor which shades down the need for working capital is that the payment of subcontractors and suppliers bills can be delayed, in some cases, for up to ninety days. In theory it is possible for the main contractor to receive payment for work in progress before it has paid many of its own subcontractors and suppliers. In addition to all this, a firm will seek to optimize its pricing structure by, so-called, forward loading. Operations which take place early in the project are priced upward with a consequent downward 'loading' of rates near the end of the project. Hence, in the early months of the project a disproportionate amount of the total contract sum is paid to the contractor. In terms of working capital the project may approach a position of being self-financing.

Finally, unlike many other industries, the product is sold before it is made. This has the implication that, with the exception of private sector housing, there is no need for firms to carry an inventory of finished products.

Structure of the UK Industry

The origins of the industry are in the primary need for shelter. In antiquity, each family built shelter for itself and groups of neighbours built structures such as, churches, bridges and flood protection works, for common needs.

Today in many developing and developed countries people still play a significant role in personally providing labour for the construction of their own dwellings. In Europe this tradition still exists in countries like Germany, France, Portugal, Greece and Ireland. It is also a common phenomenon in North America and Scandinavia. Many of these activities escape legal regulation and statistical enumeration and are carried on in what is known as the informal sector. Most informal sector work is not illegal or tax evasive. For example, housework and do-it-yourself activities are sometimes carried on by employees but mostly not.

The shadow economy consists of the informal sector and the, so-called 'black economy', the latter of which is the illegal sector. The size of the illegal sector has been estimated at 3–5% of GDP (*Gross Domestic Product*) in the UK, for instance, and less than 5% in West Germany, (Smith and Wiede-Nebbeling 1986). Informal sector activities are very important in poorer countries and provide essential requirements of society that might otherwise go unattended. For this reason the World Bank plays a role in helping the development of informal sector construction activities in many countries (World Bank 1984).

So, for quite fundamental and historical reasons, the construction industry is a key presence in the shadow economy of many countries. There is abundant anecdotal evidence to suggest the existence of illegal sector work in the construction industry of the UK in particular, (Pepinster 1989). The size of this sector is difficult to measure. It is probably safe to assume that the presence of a shadow economy does however cause greater imprecision in the quantitative descriptions of construction than many other industries, (Turner 1987).

According to the World Bank (1984: 29) the structure of the construction industry varies significantly from country to country and depends directly on three factors as follows.

First, the type of work to be done. This depends on, for example, the size of the country, its climatic and geographical features, the dispersion and density of its population.

Second, the choice of technology. This depends again on the physical and climatic conditions, the state of technological development of the country, the availability of resources, labour, materials and capital, government policies and the overall level of development of the economy.

Third, the social and economic environment. This is a function of cultural and historical conditions, the political and economic organisation of the country and the state of the economy. The structure of the organisation of production units in the industry will be consequent upon these three factors. There are usually four sets of production units, informal sector individuals, self-help or communal organisations, state owned and private sector enterprises.

A detailed quantitative description of the industry in any one country would not be resonant with the aims of this text and would become quickly out of date. Most developed countries have good statistical sources. For the UK a detailed picture of the structure of the industry and its employment and output may be had from the DOE (Department of the Environment) *Housing and Construction Statistics* published quarterly. The role played by the formal industry in the economy may usually be deduced from the national accounts of the country in question. A quick introduction to the construction industry of a country may be had from publications such as the *UN Construction Handbook* and Spon's *International Construction Costs Handbook* (Davis *et al*. 1988).

Operational Features of the Industry

Operational features of the industry evolve out of the historical, social and economic conditions but, in themselves, are far less fundamental and are capable of being changed in the short term. In the context of the UK industry we will consider aspects of the relationship between the industry and its clients.

Until the 1980s the characteristics of the relationship between the industry and its clients were largely controlled by the vested interests of the industry itself. This had advantages as well as disadvantages for the clients of the industry. The exception to this was that groups of clients who were frequent customers of the industry (banks, chain stores etc.) and thus were able to dictate the procurement arrangements. Briefly, there are three types of client/industry interface, the so-called, traditional system of contracting, management contracting and its variants, project management and construction management (often including an amount of contractors design) and finally design and build. Each of these has variants, but the three which we have chosen are representative of the range of approaches.

The 'traditional system' involves the 'client' or building owner in separate contractual relationships with a range of consultants (for example, architect, quantity surveyor, structural engineer and services engineer) and with the building contractor(s). In this system, the architect is the lead consultant. It is the job of the architect to liaise with the client and to ensure that the building is delivered on time and budget. The architect produces the design and the contractor is responsible for carrying out the architects instructions.

A significant advantage of this system is that the client has on hand an array of specialist advisors. These advisors work on a fee basis and thus there is a minimisation of any profit driven tendencies to either encourage or discourage clients from initiating more or less work to take account of changed circumstances. Although it has been pointed out on many occasions, that, for the quantity surveyor in particular, there is an interesting conflict. On the one hand he or she is responsible for the cost control of the project. On the other hand, by being paid on a fee basis, the consultant gets paid more, the more expensive the building turns out to be.

One disadvantage of this traditional system is, that it is very difficult for the client to identify who is responsible in the case of technical failure in the finished work. Secondly, the architects, found themselves in the role of managing the project. Theirs is a learned profession and they were traditionally regarded as the most senior of the construction professions. Architects are generally not trained as managers. Although creativity and the ability to 'manage' are not incompatible, it does seem, intuitively, to be the case, that people who are drawn to architecture do not often have the desire to be managers. It does not follow that just because someone is a good designer, he or she will necessarily be a poor manager, it is just that such Renaissance men and women are rare and it does seem to be an unrealistic expectation for an entire profession. Even if all architects were excellent managers, the lines of responsibility and authority under the traditional system are notoriously unclear and present immense difficulties (Higgin and Jessop 1965).

The traditional system did allow the professions to develop separate identities and to monopolise their own defined areas of expertise. This suited the professional institutes who were, quite naturally, interested in ensuring their own survival. Perhaps for these reasons the traditional system had, rightly or wrongly, by the 1970s, developed a reputation for sometimes delivering expensive buildings late.

This very reputation was among the reasons which led to the growth of alternative methods of construction procurement. The two key features of both design and build and the 'management' approaches are first, unitary responsibility to the client and second, clearer definitions of roles and responsibilities among the members of the design and construction team. By the late 1970s the UK construction industry had become much more client focused. Reasons for this change in attitude are presented in detail in Chapter nine.

Notes

[1] This categorisation into first, second and higher order goods was proposed by Menger (1981) in the late nineteenth century and has been introduced in building economics by Bon (1989: 26–7).

[2] The property market is described in detail in Balchin *et al.* (1988).

[3] The historical development of buildings as objects of exchange is lucidly described in Bon (1989: 26–9) using arguments from Hicks (1973).

Part II

Demand-side Issues

Chapter 4

Intertemporal Choices and Building Economics

Introduction

Buildings are long-lived capital assets. The period between decision and action, inception and occupation, use and obsolescence is rarely measured in months, usually in years or decades and occasionally, in centuries. More than in almost any other aspect of human activity, time is central to the design, production and use of the built environment. The passage of time is intimately connected with personal, social and environmental change. Changes in the built environment over time are one of the ways in which we gain clues to our past and its relationship with our present, notions which are crucial to our collective and individual sense of self.[1]

The time dimension has traditionally been neglected in the writing on building economics. Where it has been treated, it has frequently been in the narrow realm of comparative 'cost-in-use' or 'life cost' studies, Dell' Isola *et al.* (1981), Marshall and Ruegg (1981), Ruegg (1982), Ruegg *et al.* (1975), Flanagan and Norman (1983), Flanagan *et al.* (1989). Investment appraisal techniques which take account of the time value of money are well-known and are applied in the appraisal of construction projects, (Darlow 1980).

Three observations about this are worth making. First, investment appraisal techniques which take account of the time dimension have been adopted and used in a rather ad-hoc fashion by building professionals. As Bon (1989) has pointed out, there is a need for a full incorporation of the time dimension into a theoretical framework for the appraisal, design, construction and use of the built environment; a task which is far beyond the scope of this text.[2] Second, the techniques themselves are subject to criticisms, in particular, of their psychological assumptions. Third, recognition of the centrality and pervasiveness of the time dimension leads logically to an entirely different approach to the subject of building economics. Namely, we are forced to concede that there is little basis for deterministic plans and forecasts in an endeavour which takes place over a long time-frame in an economic environment which is itself volatile.

This chapter will proceed to consider first, the time profiles involved in building. Second, we will examine how money expenditures and receipts connected with buildings are conventionally adjusted to take account of the timing of the cash flow. Third, we will consider the nature and consequences of uncertainty about the future. Fourth, we will consider the emerging criticisms of these investment appraisal techniques. Finally, we will attempt to synthesise the issues above and to develop an overall perspective on building economics which gives recognition to the importance of time.

Time Profiles in Building Economics

The literature on 'cost planning' which grew in the UK in particular in the 1960s was, and to a large extent still is, primarily focused on the planning of capital cost. In the early 1960s the professional institutes of architecture, engineering and surveying adopted specific procedural approaches to cost planning. These are now fully incorporated into professional practice and are used by design teams to plan and control the capital cost of building. Stone (1967) was among the first to consider extensively both the initial and the running costs of buildings. Standard texts on the techniques of building economics include chapters on life cycle costing and cost-in-use techniques, Seeley (1983), Ferry and Brandon (1991).

However, the predominant perspective of building economics taken by industry professionals is still concerned primarily with the initial cost of the building, with often little more than lip-service being paid to life cycle approaches in the UK. The practice of building economics is, in this respect, more advanced in the USA where some form of discounted cash flow life cost-benefit appraisal is required for all public sector building decisions (Ruegg, 1984). Reasons for agreeing in principle but not actually carrying out life cost analysis are usually to do with lack of data on running costs, failure rates and replacement cycles; the 'impossibility' of predicting the future; arguments over rates of interest, discount and inflation and finally, the 'bird in the hand' syndrome or the tendency for decision makers to be myopic, to overvalue the present compared with the future. This lack of application in practice is rather surprising considering the increasing evidence that the time-adjusted subsequent running, occupation and functional use costs of buildings may dwarf the initial capital costs. (Ward 1987)

Total development cost includes the costs of site, site acquisition, design and construction of the building or engineering project, running and occupation costs, periodic and aperiodic repair and maintenance and the net cost or benefit of disposal or replacement of the facility. If we are concerned with the economic use of resources in the production and use of the built environment then we must take as our starting point the life cycle of the built asset.[3]

In effect, our economic model of building must encompass all of the costs and benefits extending over its economic life. Conceptually, this can be *ex-post* as an historical account or *ex-ante* as a tool for planning and making decisions among competing alternatives (Bon 1989: 13). As building economists, not historians, we are naturally concerned with the latter case. This view has been proposed by Bon (1986b, 1989) in the form illustrated in Figs 4.1 and 4.2.

> 'Consider a firm that converts a flow of inputs into a flow of outputs. We can represent the anticipated variation of these flows over time by an input–output profile, ... [Figs 4.1 and 4.2] ... The economic process consists of the construction of a plant, its operation over a period of time, and its ultimate dismantling. Here, 'plant' refers to the capital combination underlying a production plan. The process is accordingly mortal: it will have a beginning and an end. The time profile of inputs displays a 'hump' associated with construction expenditure, while maintenance and replacement expenditures are included in the subsequent inputs. If plant reconstruction is anticipated, it can be represented analogously by another such hump ...'
>
> (Bon 1989: 13)

The fact that there are difficulties in estimating the magnitude of the cost or benefit flows is not sufficient reason to ignore them completely or to confront them only in the presence of excellent information or of specific client pressure. At the end point (right hand side) of both diagrams the input and output

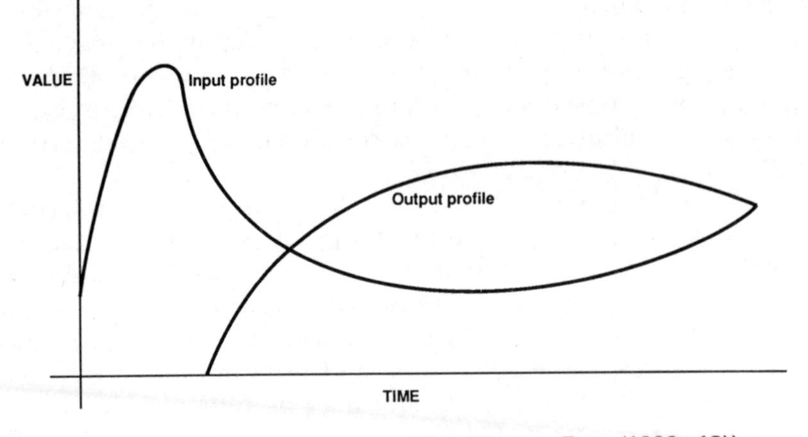

Fig. 4.1 Input–output profile. (*Source*: Bon (1989: 12))

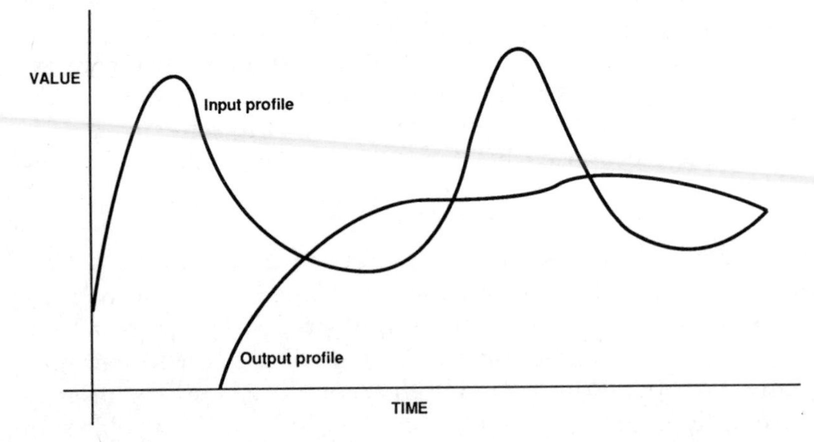

Fig. 4.2 Input–output profile with plant reconstruction. (*Source*: Bon (1989: 13))

profiles intersect. This point marks the end of the economic life of the building. The benefits (outputs) derived from the production process, of which the building is a part, no longer exceed the costs (inputs) to that process. If the building remains in use after this point it will involve a net loss to the owner. This is, of course, a theoretical optimum. In practice, the building may have a shorter or longer life. This theme is discussed more fully later.

Money and Time

Even as late as the eighteenth century, while it was quite acceptable to pay for the use of other resources, the payment of interest for the use of money was associated with usury.[4] As we have seen in previous chapters, both Marshall and the Austrians in the late nineteenth century established theoretical bases for the payment of interest which removed the latent immorality from such transactions. Interest represents a payment/cost for the use of money over time. In this section we will briefly review the principles of the techniques which are used to render equal, flows of money which take place at different points in time.

Invested today, £100 in a bank account carrying a rate of interest of, say 8%, will compound to £108 at the end of the first year. If the £108 is left in the account at the same rate of interest for a further year it will be worth £116.64 and so on. Thus, on the given information a rational decision maker would in theory be indifferent between £100 today and a certain £116.64 in two years time. This process of *compounding* can be expressed mathematically thus:

$$S = P \times (1 + i)^n \tag{1}$$

where S = compound total
P = original sum
i = rate of interest per time period
n = number of time periods (years/months etc.)

The process can be reversed. Thus, the equivalent today of £116.64 in two years time is £100. If £116.64 is *discounted* back to its present value it gives £100. This may be expressed thus:

$$P = \frac{S}{(1 + i)^n} \tag{2}$$

Therefore £100 is the *present value* of £116.64 in two years time at 8% pa.

On a similar principle, we can establish the present value of a recurring cash flow. This means for example, that we can calculate the present value of the estimated annual running cost of alternative building designs over a given number of years. This can be represented mathematically thus:

$$P = \sum_{y=1}^{n} \frac{S_y}{(1 + i)^y} \tag{3}$$

where y = time period (month/year etc.)

Traditionally among property professionals the formulae are not used, instead the multipliers are worked out and compiled into tables. The three we have referred to above are known as *'The Amount of £1'*, *'The Value of £1'* and *'The Value of £1 Per Annum'*. Using these tables it is possible to calculate the *net present value (NPV)* of cash flow profiles of alternative projects. Thus, projects which have different cash flows occurring at different points in time may be rendered comparable in terms of the net present value of their initial, running, maintenance, repair and replacement costs, considered over a defined time period. For our purposes here, a project may mean a building or a building sub-system, such as a heating system or a type of roof or window. Using this technique, future costs are 'discounted' back to the present. Hence, the generic title of *discounted cash flow (DCF)* for this type of calculation.

In projects where the majority of the flows are costs, and the associated benefits are not being explicitly considered, it is common to use the NPV as the decision criterion.[5] On the other hand, if the benefits associated with each alternative are perceived to be different and are measurable then some sort of time adjusted benefit to cost ratio, (or in cases such as energy saving retrofits), a savings to investment ratio or *internal rate of return (IRR)* would be more suitable.[6] The principle of an IRR is that having adjusted all inputs and outputs for time, it measures the percentage rate at which the future cash flows generated from the project return the initial investment. Thus, the higher the IRR the quicker the rate at which the initial investment is returned. In particular circumstances the NPV and IRR methods will yield different results. One specific example is the case of two mutually exclusive projects where project A has high initial returns and a tailing off of income in later years, and project B has a low initial return with increasing income in later years. In this case lower (single digit) discount rates will yield a higher NPV for Project B whereas Project A is likely to have a higher IRR.

Before leaving this brief overview of money and time we need to present four unresolved problems with DCF. First, the problem of inflation. Inflation affects all the future cash flows of the project. Unless there is a relative change in the prices of particular components of the project, all flows will be affected equally. Therefore, there is no necessity to take inflation or price escalation explicitly into account. If there is an expectation that there will be a change in relative prices then this can be dealt with by using different discount rates for the affected components.

Second, there is the problem of the choice of discount rate itself. This is a highly complex question to which there is no single answer. Consider first the problem of real and nominal rates of interest. If an investor borrows money at a rate of interest of say, 12% and the inflation rate is say, 5% then the effective interest rate, as far as the price levels prevailing when the repayments are being made, is approximately 7%. In this case an inflation adjusted discount rate would be 7%. However, most investors are business firms and they will in theory be aware of the average rate of return they make on their invested capital. Whether the firm borrows externally or uses finance from its own reserves, the appropriate discount rate will be the opportunity cost of the capital. Put simply, this will be the return they will have to forego from investing elsewhere in order to use the capital in this particular project. Frequently the firm will from its own records and expectations derive a 'test discount rate'. Additionally, all public sector projects are subject to a test discount rate. In the USA the Executive Office of the President, through its Office of Management and Budget, issues a standard discount rate 'to be used in evaluating the measurable costs and/ or benefits of programs (*sic*) or projects when they are distributed over time'.[7] This latter rate is currently 10% i.e., quite arbitrary.

Third, there is the problem of choosing a time period for the analysis. This in general, should not be the expected physical life of the project, it should be the expected economic life (the period over which the project is expected to provide a net benefit) or the period over which the owner is expected to take an interest in the project. The latter may be shorter than the economic life.

Fourth, there is the problem of the implicit misrepresentation contained in the terms 'costs-in-use'. An NPV or an annual equivalent is only meaningful in the context of a set of mutually exclusive alternatives. DCF calculations are carried out to enable decision makers to choose between alternative courses of action. Each NPV calculation is an *ex ante* 'snapshot' taken at a particular point in time with a particular set of information. As time passes, the information held by the economic agent changes and plans are, quite naturally, revised.

This process of planning and plan revision is a natural part of any rational economic process. *Ex ante* plan revision takes account of new information and of *Ex post* rationalisation of previous experiences. The resultant is highly subjective for two reasons.

First, *post hoc* rationalisation of events is biased due, on the one hand, to 'cognitive dissonance' or our tendency to regard past decisions as good ones even if they were not, and on the other, to the fact that in past decisions among alternatives, by definition, we are only able to observe one of the alternative outcomes. Ask ten personal computer owners which is the best PC? It is highly likely that each will answer that the best PC for their purposes happens to be the one that he or she owns. To answer otherwise would be to also publicise the fact that the respondent is not a good decision maker. Therefore the possible outcome of mutually exclusive alternative courses of action must remain as mysteries.

Second, in formulating a plan we are looking into the future and considering the available information and forecasts. This information has to be interpreted in the light of our previous experiences and our knowledge of the particular field. Hence, to regard an NPV as 'real' would be a mistake. In effect an NPV is little more than a semi-rationally counted row of red beans which we can compare with other similar rows. Therefore, the term 'cost in use' is a little misleading especially when it has implications of being the 'real' or 'total' cost of a building or building subsystem. In summary, DCF provides no more than a rationale for choice among alternatives.

Risk and Uncertainty

Economic appraisals of long lived assets are made using a variety of assumptions about the future. Some of these assumptions will be *statistical* in nature even if statistics were not used in reaching the assumption. In this type of assumption it will be possible, either intuitively or otherwise, to attach probabilities to alternative outcomes. For example, the length of time it should take to complete a particular construction operation. This is an example of a decision under risk. Alternatively, there will be situations where there is no useful past data and it is not possible to attach probabilities to alternative outcomes. These are decisions under uncertainty. Examples include the possibility of *force majeure* events such as war, and other events like unforeseen changes in Government policy. A recent example of the latter was the sudden decision by the UK Government in March 1988 to ban the use of 'sale and leaseback' schemes by local authorities. This had the effect of halting (if only temporarily) many innovative building rehabilitation schemes for public sector housing in poor urban areas which were being planned jointly by Housing Associations and Local Authorities.

Risk in economic decision making and investment appraisal is a subject in its own right which is treated in detail elsewhere, Hertz and Thomas (1983), Hayes *et al.* (1986), Cooper and Chapman (1987) for example. In the context of this discussion about events which take place at different points in time, we are concerned only with the fact that people make subjective and objective adjustments to their estimates of values of future events due to the presence of risk and uncertainty.

The Psychology of Investor's Myopia and Other Ailments

One effect of DCF as an investment appraisal method is that with high discount rates the value of cash flows in the future is diminished considerably. Combine this with the fact that as we plan further ahead, the riskiness of our assumptions increases and we quickly begin to consider only short time spans. Such 'bounded rationality' is in fact a natural way of dealing with risk by firstly considering only 'half a story' and secondly, by building-in a risk premium and only considering projects which are profitable in the short term. This is a myopic view of the future.

Daniel Bernoulli (1954; first published in 1738), one of the first writers on risk in economic decisions, and the psychologists Kahneman and Tversky (1975) have shown that people evaluate prospects by attaching subjective values to the outcomes and that there is an asymmetry between gains and losses.[8] In other words, an individual may prefer a sure gain of £90 over a 90% chance of £100, although the objective monetary expectation of the two outcomes is the same. This preference for a certain outcome over a gamble which might have a greater gain, is known as *risk aversion*. Tversky and Kahneman have developed the idea to include *loss aversion*. In other words, many people will prefer a sure gain over a gamble which has the possibility of a larger gain, and a sure loss over a gamble which may, either clear the debt or incur a larger loss. This leads to the conclusion that decision makers are influenced by the way in which choices are framed. Street (and other) market vendors of all types have long been aware of this of course. Although it has been pointed out that governments are not always aware of it, for example when urging house owners to save money by installing additional insulation they might be better served by telling people how much they are *losing* by not insulating.

Thus, investment appraisal methods which involve judgements about the future are likely to encourage short-termism and in the presence of risk and loss aversion to lead to sub-optimal outcomes which with the benefit of hindsight (or viewed *ex poste*) would be sub-optimal.

In Conclusion: Subjectivity, Time and Building Economics

Decisions about buildings take place under conditions of risk and uncertainty. There are a variety of means of dealing with these. For example, testing the sensitivity of our answers to changes in the values of particular variables, or using probability distributions *in lieu* of single values for variables. Given our awareness of these issues, to present deterministic appraisals is not only to mislead the decision maker but also to fail to make use of the full extent of our knowledge. The future is not totally predictable but neither is it totally unpredictable. The quality of our information ranges from the near certain knowledge that something will happen tomorrow or next week, through the probabilistic estimates of say, project duration backed up by statistical data, in this case, of weather records, to the highly uncertain and unknowable future.

The framework of building economics needs to be developed into one, which is about economising resources in the production and use of the built environment. It must take account of subjectivity, risk and uncertainty, by taking economic life perspectives as a starting point and by presenting all plans in some sort of probabilistic way. A well founded, theoretically consistent first step on the path to this goal s presented in Bon (1989).

Notes

[1] A detailed and beautifully presented examination of the connections between the three concepts of image, environment and individual wellbeing in the context of time is Kevin Lynch's book *What Time is This Place?* (1972). In this work Lynch develops the theme that desirable images based in the [built] environment should 'celebrate and enlarge the present while making connections with past and future.' (Lynch 1972: 1.) This is an important issue for those involved with conservation and preservation of buildings. Preserving only the best or most imposing examples of building can quickly lead to a distorted view of history. This latter point has been developed in detail more recently by Hewison (1987).

[2] An excellent start has been made on this huge problem by Bon (1989).

[3] The approach here follows Bon (1986b, 1989) insofar as he proposes an 'input-output' view of the costs and benefits of buildings over time.

[4] William Shakespeare wrote his play 'The Merchant of Venice' in 1598. Contemporary Elizabethan mores are reflected in the way he dealt with the relationship between Shylock and his clients Antonio and Bassanio. The latter pair hate Shylock for two reasons. Because he is a Jew and because he is, by profession, a lender of money, requiring collateral and charging a rate of interest. This makes him a usurer.

[5] Some decision makers prefer to calculate the *annual equivalent cost* rather than the NPV. In this case, all cash flows, including once-only flows such as plant/building purchase or replacement, are converted to their annual equivalent over the time period which is being assumed for the purposes of the analysis.

[6] The technical details of these alternative DCF methods are presented in most texts on investment appraisal, see, for example, Pilcher (1984: 16–90).

[7] Office of Management and Budget, circular number A-94 Revised, dated 27 March 1972, is reproduced in full in Ruegg (1984).

[8] This theme is lucidly developed by Bon (1986a) on which this section is based.

Chapter 5

The Demand for Construction

Introduction

The objective of this chapter is to develop an understanding of the patterns of demand for construction work generally and also for specific types of construction. First, we will define exactly what we mean by 'demand' and examine the consequences of that definition. Second, we will review the general economic and social factors which influence demand for construction. Third, we will consider the nature of demand for each of the construction sectors. Finally, we will consider the implications of the foregoing for the market in construction and property.

Demand

In economics the term 'demand' has a meaning which is not the same as its everyday definition. Demand, in the former case, refers to the amount of a good which consumers are both *able* and *willing*, as apposed to merely wishing, to purchase at each price in the conceivable range. This relationship between price and demand may be described in terms of a, so-called, demand schedule. A demand schedule should show, in numeric terms, the amount of a good which will be demanded at each discrete price in the range. When this relationship is plotted, the result is the familiar downward sloping demand curve. Thus, assuming a rational utility-maximising consumer, we conclude that, other things being equal, if the price of a good increases then the quantity demanded will fall. It is worth emphasising that, in this case, there has been no change in the underlying conditions of *demand*, we have merely moved along to another point on the demand curve. When factors other than price change, we may find ourselves on a new demand curve. For example, if the consumer's income increases, then he or she will be able to afford more of the goods at any price level in the range. The demand curve will have moved upwards and outwards.

A consumer's decision to purchase is influenced by many things. Consider for example the following; the weekly family trip to the supermarket for food and provisions, the impulsive purchase of a paperback novel at a railway station bookstall, the, in the UK at least, long drawn out negotiations to purchase a house or flat, the decision to take a loved one to the theatre followed by a late meal at an expensive restaurant. This series of purchases demonstrates complex motives including, the need to relieve hunger, the need for shelter, the need to relieve boredom and the desire to seek pleasure and stimulation. If the price of popular fiction falls, will the impulse railway station consumer necessarily increase the quantity demanded, i.e., buy two novels instead of one, three instead of two, seven instead of six? How many novels can you read on a two hour train journey? Or, is the quantity of novels demanded related to the waiting time spent on the platform? Clearly, in considering only the relationship between demand and *price*, we are using a very simplified model of human behaviour.

Accepting then, for the time being, this rather one dimensional view of the rational profit/utility maximising decision maker, we need to introduce one other feature of demand, namely its price sensitivity or elasticity. The responsiveness to price changes of the quantity demanded is known as the *price elasticity of demand*. If the demand for a good is very sensitive to price the demand for a good is said to be elastic. Conversely, inelastic demand is where quantity demanded changes little when price changes. The price of bread, for example, is relatively inelastic, people have to eat. If the price goes up, then consumers will allocate their resources in different ways in order to be able to buy the bread they need to feed their families. Luxury goods are said to be elastic. If the price goes up, other things being equal, people will buy less. If the price goes down, there will be a tendency for people to buy more, within a certain range.[1]

Finally, as Hillebrandt (1984: 9–10) has pointed out, it is easy, in the context of the built environment, to confuse 'demand' with 'need'. For demand to exist, someone must be both willing and able to pay. There are, for example, thousands of families living in squalid temporary hostel accommodation in the cities of the developed world. This points to an immediate 'need' for habitable, permanent dwelling units. Many local authorities and public sector housing associations are *willing* but, unfortunately, not able to pay for the needed units. Thus, these needs do not get transformed into demand. This is not a trivial distinction between need and demand. In order to understand the demand for construction we need to understand, not only how the built environment fits into and serves the needs of society, but also, how people and firms are empowered to commission and pay for buildings by being able to raise relatively large amounts of money via the economic system.

The Economic and Social Environment for Demand

In broad terms the demand schedule for any commodity is primarily derived from five interrelated, and, to some extent, interdependent factors. These are as follows. First, the price of the good. Second, the real income of the purchaser. Clearly, rich people are in a position to buy more of everything should they so want. At lower levels of income there is no choice. Third, the quantity demanded is affected by the size of the market. Other things being equal, a doubling in the number of family units should double the demand for dwellings, television sets and so on. Fourth, the demand schedule will be influenced by the price and availability of close substitutes. Many people living in London find it unnecessary to own a car given the price and availability of tubes and buses. On the other hand, in suburban and rural areas there may be far less public transport and thus no close substitute for car ownership.

Finally, demand is influenced by peoples' tastes and preferences. There are many arguments as to whether peoples' tastes and preferences are affected by advertising or not. It seems undoubtedly true that tastes and preferences are formed by a complex array of personal, social and exogenous forces of which advertising is one. Thus, hidden and not so hidden, persuaders can influence tastes which in turn influence demand. Thus, it is possible that this particular factor affecting demand may be subject to influence by firms. In fact, if it is not there is little justification for the multi-million pound advertising industry.

Before proceeding to the details of particular sectors it is appropriate to sketch in the general context for demand, provided by the prevailing economic conditions. Conventionally, when GDP is growing at an acceptable rate, when inflation, interest rates, and unemployment are low, conditions are regarded as being favourable for investment. Buildings are investment goods. Housing though, is regarded largely as a consumption good. Given the long time-spans involved in investments in construction projects, it is at least as important that there are *expectations* that these conditions will arrive/continue.

It is also worth mentioning that when we refer to interest rates we are referring to real, not nominal, rates of interest. In the UK at least, housing is an exception to this. The statistics for private sector output in the 1980s indicate that housing production may be more directly related to disposable income than to real interest rates. Although in reality this may not be very important as the prevailing interest rate merely converts disposable income into effective demand in a highly geared fashion.

Focusing now on the construction sector, the demand for built space and for other constructed facilities depends on a large number of interrelated economic and social factors.[2] First, the demand for construction is profoundly influenced by the size, the structure and the geographical distribution of the population. Thus, the age structure of the population, the average family size and the patterns of migration are all very important, not just for housing production, but also for other social facilities and amenities such as, schools, hospitals, libraries, sports centres, roads and railways.

Second, the demand for construction is influenced by the condition of the existing stock. The existing stock here refers not only to buildings as such, but also, to sewage and drainage systems and other elements of the infrastructure.

Third, the question of suitability of the existing stock will depend in part, on the climate and on how heavily the facility has been used, and in part on how technology has changed. For example, how many buildings in the contemporary rural landscape are part of the technology of modern agricultural production and would this have been unimaginable fifty years ago? Nineteenth century cowsheds are, on the whole, not suitable for large milking machines. Similarly, industrial production has moved from production lines to robotics, office work from paper to telecommunications, each having major consequences for the buildings to house the process. Notions of technological or economic life, as opposed to structural life, are explored in detail in Chapter six.

Finally, one overall implication of the factors described above is that the demand for construction is frequently subject to large fluctuations. This is a tendency inherent in capital goods industries where relatively small changes in demand by consumers cause large 'stepwise' expansions or contractions in production capacity (World Bank 1984).[3] This holds true in most developed and developing countries. Cyclical variations of relatively large amplitude in output and, by implication, demand, lead to inefficient use of resources over time. Consider just one factor of production, labour, for example. In times of recession, workers are laid off, sometimes never to return. This net loss of skills becomes an issue when the cycle swings the other way and skilled people are needed. Training takes time, four years or so for a craftsperson and six to eight years for a professional engineer or architect. A study by the World Bank (1984) showed that output in construction varied more than either manufacturing or GDP for the following countries; Brazil, Colombia, Ethiopia, Federal Republic of Germany, Ghana, Italy, Japan, Kenya, Liberia, Malaysia, Peru, Sri Lanka, Sweden, UK, USA and Zambia. Although it should be noted that there are two empirical studies which do not support this thesis for the UK (Sugden 1980; National Economic Development Office 1976.)

It could be argued that because such large variations in demand lead to inefficient use of resources, a policy of demand management needs to be implemented in order to smooth out these peaks and troughs and to facilitate steady growth and effective planning of the capacity of the industry. The governments of many countries are ideally positioned to do this, being responsible for macroeconomic management on the one hand and for commissioning large portions of public sector output on the other. The proportion of output which is commissioned directly or indirectly through the public sector varies with the stage of development of the country, from around 80% in less developed countries to around 40% in places like Central Europe and North America.

Demand by Sector

It is convenient to consider the demand for construction in the sectors used by the Government for statistical data collection. The determinants of demand have been described in detail by Hillebrandt (1984: 9–38), the following paragraphs are loosely based on her account.

Public sector housing

Traditionally, public sector housing for rent was provided for families who were not able to afford to obtain housing of an acceptable standard in the owner-occupied sector. Although in the UK public sector housing was a 'general needs tenure' in the 1920s and again from 1945 to the late 1970s. In the period 1979–1988 the output of public sector housing in the UK fell dramatically. One of the main reasons for this was the policy of central government which favoured reductions in the size of the public sector. This demonstrates well the case that although public sector housing in the UK had become known colloquially as 'Local Authority' housing, real control over initiating and financing projects had increasingly gone over to the central government. It should be noted that the government's 'right to buy' policy for occupiers of public sector housing enjoyed much popular support.

Public sector non-housing

This sector divides broadly into two types of work. Firstly, projects for the 'semi-state' bodies such as coal, gas, electricity and water boards and secondly the provision of the built infrastructure of roads, harbours, hospitals, libraries, schools etc. This sector is largely controlled by central government through their forward plans for public expenditure on capital projects.[4]

Owner-occupied housing

Demand for private sector new housing depends on the size and condition of the existing stock, demographic change, real interest rates, and the level of disposable income of the family unit.

Private sector industrial and commercial work

Demand for this type of construction is dependent on the demand for goods and services in the economy as a whole and is thus known as 'derived demand'. There may be large regional differences in the level of derived demand.

Rehabilitation, improvement, repair and maintenance

Demand for this type of work will be related to the estate management policy of the appropriate authority, the condition of the stock, the level of perceived obsolescence. Government expenditure plans will again directly control the public sector while the general economic conditions will affect private sector plans. It is worth adding that government expenditure policy will also significantly affect private sector plans through the availability of improvement grants and other incentives.

The Market for Construction

Perhaps unsurprisingly, the construction industry has a highly fragmented structure. Construction work may be segmented in three ways; by size of the firm, geographical area and by the type of work.

There is a large number of small firms. Approximately 90% of the firms employ fewer than eight people. The 100 biggest firms in the country account for just under 20% of the industry's output. People working in the industry commonly divide contractors up into so-called 'small', 'medium' and 'large' firms. There is no universally agreed set of criteria to define exactly what is a small, medium or large firm. Nevertheless it does seem intuitively correct to differentiate between even fuzzily defined groups of firms. It is tempting to define the groups by the size of project they undertake. This could suggest that, say, small firms carry out projects up to £x in value, medium firms up to, say, £2x and large firms above, say, £4x.

This categorisation has a major drawback in that it is possible to envisage a firm which undertakes only very small projects, but very many of them. Is this a small or a large firm? Therefore, it seems intuitively more helpful to categorise firms in terms of their turnover or organisation and management. Thus, a small firm would not have any separately identifiable 'management' personnel. A 'medium' sized firm would operate with a larger turnover probably in a larger geographical area with management personnel who may also carry some executive responsibilities. Using these loose criteria, a 'large' firm would probably be operating nationally and internationally, involved in multi million pound projects and with a clear differentiation between executive and management responsibilities.

Thus, for a given project in a given geographical area, the number of firms who would be interested in, or capable of, undertaking a given project may be quite limited. For example, large national firms may find that they cannot compete with medium sized locally established firms for medium sized projects unless they are willing to tender for the project at a loss in order to get established in a new area. Similarly, small firms may find that travel and subsistence costs prohibit them from expanding into new geographical areas when locally established firms do not have these additional costs. One answer to this is to set up a new administrative office and then to recruit local labour.

Similarly, firms tend to have some specialisation in the type of work they undertake. The firms in the industry may be divided into building and civil engineering, and further into general contracting, housebuilding, repair and maintenance and specialist subcontracting. Specialist subcontractors (for example; ceiling, flooring, services subcontractors) undertake work in many market sectors and geographical areas. Thus the perfect market criterion of a large number of suppliers is frequently not met. Also there is rarely 'perfect knowledge' due mainly to the tendering methods outlined later. However, there is 'ease of entry to the market', as the working capital requirements of construction firms are relatively low.

In summary, there is no 'market in construction', however there is a series of rather imperfect 'mini-markets'.

Notes

[1] This is an oversimplified presentation of the oversimplified concept of demand. Among the features we have not dealt with here, is the aggregation of individual consumer demands into market demand. The reader is referred to Samuelson and Nordhaus (1985: 377–432) and Mulligan (1989: 15–53).

[2] These factors have been clearly presented in some detail by Stone (1983: 227–44) in an additional new section in the third edition of his *Building Economy*.

[3] This notion of 'stepwise' changes in production is developed in more detail in Chapters 8 and 9.

[4] Between 1972 and 1986 the output of public-sector work not including housing in the UK, fell by 40%. It appears that the bulk of the decline was in the building of roads, bridges, educational buildings, water mains and sewers Briscoe (1988: 126–30). Many British motorways were by then well over 20 years old, in need of major repair, and in some cases redesign, to accommodate much larger traffic loads.

There was a major economic policy problem here for the Government. On the one hand they were trying desperately to control inflation using their chosen tools of interest rates and public expenditure controls. On the other hand they had contributed to the heightened demand by their tax cuts of the 1988 budget. The expenditure needed for renewal of the built infrastructure would of course, via the accelerator, increase demand and inflation even more in the absence of tax increases and/or credit controls. However, tax increases and credit controls involve intervention in the economy to an extent which the government found unacceptable.

Chapter 6

Building Obsolescence

Introduction

Just under half of the output of the construction industry of the UK, and many other developed countries, is produced by work which takes place in existing buildings. The latter are obsolete in some way and in need of repair, maintenance, rehabilitation, adaptation or improvement work. It has been said that obsolescence is the cause to which rehabilitation is the response. Obsolescence, decay and renewal are three points on a continuous cycle which pervades the built environment.

Obsolescence in the built environment is inextricably linked to change and is usually a symptom of some other social, technological or economic change. The life histories of buildings are diverse. Between the two crucial points of construction and demolition lies a varied pattern of existence where buildings are subject to periods of occupancy, vacancy, maintenance, modification, extension, until their eventual removal by demolition. Within this large and cumbersome system, prone to influence by many external agents, there are patterns. Buildings are planned, constructed, used, adapted, reused and demolished. The end of the life span is marked very definitely by demolition, less definitely by obsolescence and sometimes, but not always, by both at the same time. Demolition is a clearly visible state. Obsolescence is a more complex state and is not easily defined. In this chapter we will proceed first, by considering the nature of change in the built environment. Second, we will discuss the nature of obsolescence in the built environment. Third, we will present the critical factors which affect the life of individual buildings.

Change in the Built Environment

There are connections between individual well being, social change and change in the natural and built environment. Seasons mark the passage of years, and birth, puberty, adulthood, senility and death, mark human growth and development. These indicate both natural and human biological rhythms to time. Just as we can observe change in nature, we can mark the passage of time by changes in the built environment. Change, growth, development, decay and death or renewal, is as inexorable in the built environment as it is in nature and in ourselves. We gain clues as to the passage of time by making observations of our environment; light and dark, indicating day or night; traffic jams, crowded cafes, opening and closing of bars, cinemas and theatres indicating the (urban) time of day. Lack of access to the data on which to make these observations diminishes us. Windowless rooms with permanent artificial lighting are disorientating. Prisoners are placed in solitary confinement as a punishment.

What is the relationship between change in the built environment and social change?[1] In general, change in the built environment is a response. For example, changes in the technology of newspaper production, telecommunications and the opportunity cost of office space in the early 1980s caused many buildings in Fleet Street and in the City of London to become unsuitable for newspaper production and financial services. One direct result of this was the rapid transformation of large areas of derelict land in the London docklands into an up-to-date centre for newspaper production and financial services using contemporary technology. Changes in the built environment may be read as indicators of social change.

Examples of this include:

(1) Major programmes of high-rise public sector housing, built in response to housing crises in the UK in the 1960s and, for example, in the Ursynow and Natolin suburbs of Warsaw in the 1980s.
(2) Inner city decay in, for example, Liverpool, Manchester and Birmingham as the residents and industries moved out to the suburbs and to locations better served by motorway links respectively.
(3) The emergence, in the USA in the 1970s and the UK in the 1980s, of very large, out-of-town, shopping malls in response to the increased level of automobile ownership and changes in the work, social and leisure patterns of affluent families.

Examples of change in the built environment causing change in society are less frequent but usually very traumatic. They arise from precipitate, major transformation in the built environment such as are caused by disasters or migration. The fire which started in a Pudding Lane bakery on Sunday, 2 September 1666, burned out of control for four full days, left 80 000 people homeless and the major part of the city of London in ashes. By 1673 most of the reconstruction was complete, a major achievement by the standards of the day. The new streets were wider, the new houses were made of brick. However, many of the poor people could not afford the new houses and moved permanently to new outer villages. Many of the merchants and traders found the lower taxes of the suburbs to their liking. The demise of the vast power of the old Guilds was hastened by the necessity, against strong opposition from those same Guilds, of bringing in rural tradesmen to ensure the progress of the construction. The social fabric of the city was changed irreversibly.

Contemporary examples of traumatic major change in the built environment which may cause change in society include, the Mexican earthquake of 1986 and the Armenian earthquake of 1988. The latter left 55 000 people dead, 500 000 homeless and the city of Leninakkan virtually razed to the ground. Ironically, this terrible human toll was primarily caused by the collapse of badly designed and badly built modern structures. It is still too soon to assess the full consequences of these tragedies for the surviving society.

Obsolescence in the Built Environment

The word obsolescence comes from a Latin verb which means 'to grow old'. While the etymology contains a grain of truth, it also contains the seeds of misconception. Obsolescence is not the same as ageing. Just as it is possible for an old building to be useful, it is also possible for a new or recent building to become useless. In the late 1980s, examples of recent buildings which were obsolete included many high-rise apartment blocks and city centre shopping precincts built in the 1960s and 1970s.

Two factors combine to make obsolescence in buildings both different and difficult to study. First, buildings have relatively long physical lives, there is more time for external conditions and our requirements of buildings to change. A building with a planned physical life of say, sixty years can become technically obsolete before half its life has passed. For instance, the three decades up to the 1950s saw, in the food processing and many other industries, the introduction of continuous material flow systems, automatic controls and ovens with moving floors running hundreds of feet in a straight line. In industry, at least, the shape and nature of the building is often determined by the process. New advances are being made in manufacturing technology and robotics in elapsed times far shorter than the physical life of the flimsiest building. The technical life of the building will always be shorter than the physical life in these cases. In reality the picture is much more complex as a number of different life cycles for building components will be superimposed by the terms of leases, by changes in technology and in market conditions. For example, structures may be designed for 50–60 years, major services for 15–20 years, finishes and fittings 3–10 years, IT installations 2–3 years and so on.

Second, in addition to being fixed in time, buildings are fixed in location, they cannot adapt to their environment by moving around. This means that, to the usual factors determining demand, such as supply, quality, utility, price, availability of substitutes and so on, we must add the fact that buildings are located in a spatial market. From the point of view of a consumer, it is quite possible that the building most suited to the process he or she wishes to carry out, is situated in the wrong place and so is of no practical use.

These traits give rise to a number of unique features of building life. Buildings are relatively expensive, requiring high levels of capital for initial construction. Although almost all buildings are designed by professional designers for specific functions, many of them are in second-hand use and are being used for some purpose other than that for which they were originally designed. Due to the locational nature of the market for buildings, demand for them is influenced by movements in social and economic activity. Innovations in transport and communications also affect the desirability of buildings in their given locations. The fact that buildings have long lives accentuates the consequences of these factors and increases the probability that, at some stage, they will be significantly affected by some external change.[2]

Disuse and antiquation are symptoms, not causes, of obsolescence. We may say that obsolescence is a continuous process manifested by decreasing benefits arising from the use of the building. Obsolescence then, is the transition towards the state of being obsolete. When the cost of providing the benefit or of adapting the building to provide the benefit, is greater than the benefit obtained, then the building may be said to be obsolete.

Remember though, that rational, consistent measures are extremely difficult to produce and are, inevitably, subjective. The key factor is change with which the building cannot cope. Many buildings become obsolete merely because other newer buildings can do the same thing better, at the same cost or less. Buildings can survive relatively long periods of obsolescence before coming obsolete, due to the fact that they are periodically renovated, adapted, and extended. Consequently, it may be difficult to establish the actual life of a building as it may have been changed out of all recognition. Evidence of this type of problem was found as long ago as 1917 in a study of railway station buildings in the USA (ASCE 1917).

Some people say that there is an academic cottage-industry in identifying, ever more subtle, overlapping, new forms of obsolescence. Our intention here is not pedantry but many of these categorisations, while they are not mutually exclusive and of little use for taxonomy or data collection, do shed some additional light on built environment obsolescence and are worth reviewing briefly.

According to Nutt *et al.* (1976) buildings can only be truly defined as obsolete when they have become completely useless with respect to all possible uses they may have been called upon to support. A key concept is the 'degree of uselessness' of a building relative to the conditions prevailing in the property market as a whole. This question of degree may be determined by the occupiers of the building, the market, or the planner as the case may be. So it follows that we are discussing a rather subjective and relative term.

Kirwan and Martin (1972) tried to formulate a more precise definition by splitting the concept into 'Physical obsolescence' and 'Economic obsolescence'. They define the former as being simply the deterioration of the physical structure of the building asserting that it is not simply related to the age of the building but rather to the joint product of the age and use of the building and the nature and scale of the maintenance that has been carried out. Economic obsolescence is defined as 'when the benefits less the costs of continuing to use the building in its present state are less than the benefits, less the costs (including renewal costs) of using the building or site for some alternative purpose' (Kirwan and Martin 1972: 20). There is not necessarily a relationship between these two. Physical obsolescence may be a cause or consequence of economic obsolescence. This definition of economic obsolescence has the advantage that it is such that we can put figures to it and so is useful for appraisal of projects.

As more work was carried out, the identification of factors inducing obsolescence has led to a confusingly large number of categorisations. Functional and locational obsolescence ensues when the building becomes unable to support efficiently the activities it contains (Cowan 1965). Site obsolescence exists when the potential value of the site becomes higher than that of the building and high enough to justify demolition and redevelopment. On a wider scale, environmental obsolescence of a whole neighbourhood may occur when the conditions there render it increasingly unfit for its current use (Medhurst and Lewis 1969). More nebulous concepts include 'style obsolescence' where a group assesses the worth of a building in terms of its stylistic qualities and 'control obsolescence' where the regulations and laws that govern planning and construction have the effect of inducing obsolescence (Alexander 1965).

This motley group of ideas is the result of researchers working individually in many different countries, clearly some rationalisation is called for. It will be beneficial to look in greater detail at the historical development of the theories and to attempt to sum the parts into a coherent whole.

Theories of Obsolescence

Historically, the study of obsolescence has gone through three distinct stages. At the beginning of this century obsolescence was viewed essentially as a process of physical deterioration, this is the most straightforward form to identify and measure. Post-war studies viewed it as mainly an economic phenomenon by studying the economic life, cycles of investment, costs in use and returns. The more recent work has tended to examine the changes and developments which take place within the organisations occupying buildings as they move from one building to another, noting when and why certain types of buildings and location become obsolete for a particular organisation.

Kirwan and Martin (1972) developed a theory in terms of the lack of maintenance expenditure needed to keep the building in an unchanged physical state. Obsolescence occurs:

> 'when the net benefit that can be derived from undertaking the maintenance necessary to keep the condition of a building unchanged or from renewing it, is less than the net benefit of

allowing it to continue to be used in its present state or for its present purpose without any (or adequate) expenditure on maintenance'.

(Kirwan and Martin 1972: 22)

From this type of theoretical study it has been suggested that the life of an asset depends on the standard of maintenance, and that the lives of many assets can be almost indefinitely extended by the replacement of parts. This view is challenged here. The life of a building is certainly influenced by the standard of maintenance, but only the physical life can be infinitely extended by the replacement of parts or sub-systems. The physical life now has such little bearing on the actual life that in most cases (a major exception being housing) the maintenance effect will be unimportant. One other exception to this is where a tenant is nearing the end of a lease, he or she may not consider it to be worth while to carry out certain items of maintenance which are capable of being postponed even under a 'full repairing' lease. Over time this may have the effect of showing a noticeable increase in physical obsolescence in buildings which were already on the way to becoming obsolete anyway. If a number of adjacent leases are due to expire around the same year, there may be serious physical decay in the entire neighbourhood.

In an attempt to weave a fabric from these strands of theory, we will note three approaches. First, building decay, functional obsolescence and environmental decay are closely interwoven. Obsolescence is a function of human perception and decision. Hence it is possible to categorise types of obsolescence based on the identity of the person making the assessment (Medhurst and Lewis 1969, Lewis 1979). For example, tenant or functional obsolescence occurs when the tenant no longer considers that continued occupation is his or her best course. Rental obsolescence occurs when the landlord feels that the existing rent agreement is out of date. This may be 'upward' or 'downward' depending on whether the rent should be raised or lowered. Clearly, no landlord would willingly decrease the rent. Downward rental obsolescence occurs when the prospective tenant can get either a better building for the same rent or an equivalent building for a lower rent. In order to avoid vacant space, the landlord in these circumstances will need to reduce the rent. Condition obsolescence occurs when the landlord feels it would be profitable to spend some money on the building either in order to maintain the present level of rent or to gain a higher rent. Tenant or landlord obsolescence can be brought on by rental or condition or building obsolescence. According to Lewis (1979: 135–41) building obsolescence is where the landlord feels the most profitable course is to demolish the building and re-develop the site for another purpose. This can only occur when rental and condition obsolescence have been examined and found less profitable. The usefulness of developing such a system of minute categorisations should be questioned but the general suggestion that the level of obsolescence varies in part with the assessor be they landlord, tenant or owner occupier is useful. This is important. It marks the fact that subjectivity is important in describing obsolescence. Furthermore, it fits in very well with the generally accepted theories of value discussed in detail in chapter two.

Second, it is possible to view obsolescence in terms of the interface between the building and the human organisations it contains and supports. (Cowan and Sears, 1966). Investigating the changing size and function of organisations, it is possible to view the development of functional obsolescence as the increasing misfit in time between the activity requirements and the building provisions. Thus it becomes possible to produce two linear scales, one for buildings and one for organisations which can be matched in order to measure degree of misfit of particular buildings for particular organisations. There are a wide variety of building types, some highly flexible and suited to any one of a number of activities. Buildings could be placed on a linear scale varying from specialist to non-specialist, at one pole would be concentrated the multi-purpose buildings such as certain types of factory and office building at the other would be specialist buildings such as, for example, hospitals and bus-stations.

Similarly, from the point of view of occupiers, some activities can take place in many different types of building, whereas others are specialised and have special requirements and thus can only be carried out in particular types of building. For some types of activity, the building is most important of all, for others it is the location of the building which is crucial. Again, we can place the set of activities along a linear scale placing the ones with specialised requirements at one pole and those without at the other.

Clearly, when these two scales are matched we can note the compatibility of particular types of activity with particular types of accommodation. Although this does serve as a very good starting point for a study of the obsolescence of the building stock as a whole, it does have certain limitations. Namely, it matches only items of activity with items of accommodation, taking no account of the interdependencies of certain activities and of the spatial interrelationships among elements of the urban fabric.

Finally, it is possible to take the view that a building offers an array of resources to a potential occupier or to an actual occupier. These resources are of two types, those contained within the building and those potentially available within the locality, each type of resource for a particular building will have qualitative and quantitative features. The physical resources offered by the building itself will include, floor area, internal division arrangements, standard of services, structural condition and floor loading capacities. The locational resources include the availability of transport, personnel, materials, customers, leisure facilities and the co-location of other similar industries as for example in 'Silicon Valley, California' and in the 'Thames Valley' in the UK. The combination of these will give a financial profile for the various uses to which the building can be put.

Conversely, it is possible to assess the resource requirements of the organisation in occupation or the potential occupier. So a theory must account for the combination of change that leads, over time, to increasing amounts of resource imbalance at the interface between building and occupier giving rise to disutility, dissatisfaction or misfits. From the point of view of the building, only one thing can lessen the effects of this resource imbalance and that is adaptability.

Measuring Obsolescence

In the final analysis obsolescence can be measured in terms of the (real or nominal) decrease in a building value. This is more commonly known as depreciation. A *decrease* in value is not always apparent. For example negative aspects of depreciation in value due to technological obsolescence may well be outweighed by increases in value due to market conditions. Unlike many other assets, buildings are attached to land and exist in a complex legal and institutional framework of leases and tenancies. Thus, at a general level, there are three potential causes of obsolescence. First, the onset of physical, technological or some other form of obsolescence germane to the building itself. Second, changes in the value of land. Third, depreciation of the leasehold interest in the property.[3]

Depreciation is very difficult to study empirically. A decline in building utility for some reason or other, will only be translated into a reduction in rent or market value if there is zero inflation and if the market is in equilibrium. To take an example, a building suffering from obsolescence may well find a purchaser/tenant at a price/rent close to that obtainable for a new building in a market/location where there is an undersupply of space. Conversely if the market is in a state of oversupply then purchaser/tenant can afford to be particular and inferior buildings will remain empty.

Salway (1986) in a seminal study of depreciation compiled valuation data for a set of hypothetical office and industrial buildings in, respectively, thirty two and twenty five locations in the UK, see Table 6.1.

Table 6.1 Rental value as a percentage of the rental value of a new building.

	Office	*Industrial*
New building	100 (−)	100 (−)
5 years old	85 (5.5)	86 (5.0)
10 years old	72 (8.7)	71 (7.6)
20 years old	55 (9.1)	52 (10.8)

Source: *Salway* (1986: 66)

Note: The figures are averages. The figures in brackets are the standard deviation.

The valuations for Salway's study were all carried out in June 1985 and were, therefore, net of any inflation effects. The problem of market disequilibrium remains though as it is impossible to quantify what effect this could have had on the results. The broad trend indicated by these figures is that, other things being equal, a 20 year old building is valued at just over half the value of an equivalent new building. This is the type of information needed by property owners to establish the optimum point for intervention by rehabilitation/demolition to increase the value of the property.

Factors Affecting the Life of Buildings

Adaptability

In the majority of cases where the physical life is longer than the functional life the amount of adaptation which can economically be carried out within a building becomes an important determinant of the actual building life. For example, if because of changing requirements an occupier needs a new type of layout or a different form of building to the existing, then there are two courses of action.

(1) He or she may sell or demolish the existing building and redevelop either on the existing site or on some other.
(2) Either through good fortune or foresight in design, the building may be such that it can be adapted to the new requirements at an economic cost.

If the occupier takes the first course and elects to sell, then virtually the same decision process will have to be carried out by all potential buyers. It is possible that the building will begin the process of filtering down through less remunerative uses. This, coupled with the physical neglect which so often accompanies less profitable uses, will normally accelerate the process of obsolescence of the building, constituting an effective truncation of the building's potential life.

Adaptability is required at the onset of functional obsolescence. In the short term, the latter may be warded off by flexibility of the internal layout where furniture, fittings or machinery can be relocated to suit small changes in requirements. Achieving the same result in the longer term may require major alterations and adaptation of the existing structure, services and internal division. The above may be rationalised by identifying three types of change which cause functional obsolescence, these are:

- Use change.
- Technological change.
- Change in standards.

The broad principles governing adaptability and the reasons for its provision are embodied in the concept of 'Long life, loose fit, low energy' as outlined by Gordon (1974). The physical reality of the provisions for adaptability in buildings may be approached under three headings, namely:

- Structure.
- Services.
- Internal design.

Structure

Structural adaptability may be necessary in order to accommodate changes in services or internal design. Primarily, structural adaptability is necessary to cope with changing requirements regarding the nature and the amount of space provided. In the former case, structural adaptability is necessary to ensure that floors and walls/frames can be adapted to cope with changes in floor loading which may be needed for new equipment. There are two ways of achieving the latter aim.

(1) *Horizontal extension*
This is usually provided for by means of a basic shell. If a repetitive structural form is used (i.e. some form of standard grid frame), then extension is possible on all four sides of the existing building within the confines of the site. A repetitive building form is a more basic version of this where the process is carried out in buildings which are standard units, as production increases more units are built. Horizontal extension in two directions can be provided for by having a building which has a standard cross-section which is then capable of extension at either end. The common factor to all these being, of course, that some part of the site must be left unbuilt on. In many cases this will not be a viable solution because of high demand for land.
(2) *Vertical extension*
This may be provided for by over-design of the initial structure. Vertical extension is generally uneconomic where provision has not been made for it at the design stage. Ironically, although this is a solution to the provision of adaptability, where demand necessitates that full use be made of the site initially, such sites will often be governed by stringent plot ratios and planning regulations which may well prohibit vertical extension from being considered as an option.

Services

Services can usefully be regarded as subsystems within the overall system of the whole building. Service systems usually have shorter lives than the buildings which they occupy, both because of their physical nature and the fact that they, in particular, are prone to the increasing rate of technological obsolescence, left in the wake of advances in science and technology. It is significant that modern buildings contain an increasing amount of service installations which in capital cost terms is often equivalent to 33% or more of construction costs.

Provisions for adaptability in design and assessment of the degree of adaptability in existing buildings for the purpose of estimating their useful lives, demands some degree ot certainty about future developments. It is in the field of services, both engineering and, more particularly, data processing and telecommunications, in which there is the most rapid technological development which must therefore be accompanied by uncertainty as to future states. So much is this the case, that it would seem reasonable to suggest that, in the foreseeable future, service obsolescence may become the single most important determinant of the useful lives of buildings.

Internal design

Whereas structural adaptability is normally a medium to long term requirement, adaptability of the internal design is needed in both the long and the short term. Relatively minor changes in the process or the structure of the organisation can have repercussions upon the layout of the building. Adaptability is achieved through the furniture, fittings and internal partitions.

A high degree of adaptability in internal design is offered by the use of demountable partitions, but these are normally relatively expensive in terms of initial costs and thus are not usually installed unless the designer can predict future changes with a high degree of certainty. Non-loadbearing fixed partitions lend themselves to adaptation of internal design but incur more disruption and cost at the time of change. Loadbearing fixed partitions offer the cheapest initial solution due to savings in the size of structural members but then place an effective veto on any future adaptability. The increasing popularity of open plan solutions particularly in office buildings can be attributed both to fashion and to the fact that flexibility of the internal layout can be achieved by movement of full size partitions. Open planning has been used very successfully in office buildings for letting, where the incoming tenant can erect a system of partitioning to suit his or her own unique requirements.

Finally, any real assessment of adaptability must take into account the interrelationship of the different parts. The structure has implications on the services and internal design and so on through all the permutations of elements. Consideration must be given to the physical lives and predicted frequency of change of all the elements and components. In this way, an order of priorities for adaptability can be derived for a particular organisation or process or function of a building.

Any decision to incorporate a degree of adaptability, by whatever means, at the design stage incurs an immediate marginal cost in excess of the base cost necessary to provide the space required to meet the immediate functional needs. This is a choice between making an initial known expenditure now or an uncertain one at some stage in the future. The degree of adaptability actually incorporated as a result of a conscious decision depends on the process the client requires to be carried out within the building, his or her financial position, the period likely to elapse before change is needed, the type of change predicted and the degree of certainty with which all the predictions can be made. Ironically, if the building is initiated in a milieu of circumstances liable to frequent change where high adaptability would be important, then there is a high probability that in anything other than the short term the change will be largely unpredictable. Many building owners in this situation tend to take the cautious accounting option of writing off the building over a short period and then they hope to enjoy a period of surplus rent at the end of amortisation. If this period of surplus rent does not in fact materialise then they would have no worries about selling or demolishing the building as it will have paid for itself anyway. Logic of this sort lends increased weight to the argument that functional/economic life is the major determinant of actual life.

In Britain a report published by the Building and Civil Engineering Economic Development Committees put forward the view that much of British industry was hampered by poor buildings unsuitable for modern production methods and that, although age was not itself a problem, many buildings designed in the previous ten years had been rendered obsolete as a result of technical change in the industries concerned (Groome 1978). Problems discovered in factory buildings included the following:

- Insufficient headroom.
- Close column spacing.

- Poor manouverability.
- Too many levels.
- Inadequate loadbearing capacities.
- High heating costs.
- Difficulty in keeping surface areas clean.
- Unsuitable arrangements for delivery and dispatch.
- Insufficient access for handling goods.
- Unsuitable staff facilities.

It is significant also that the response to the survey which formed part of the research was more than 60% better than was expected, thus indicating that this was a subject of real interest to managers.

In a more recent study Salway (1986) interviewed 63 property investment managers who, collectively, managed £21 billion worth of property. The managers were asked to rank six problems often associated with older buildings. A points system (3 points for a first, 2 for a second, 1 for a third) was applied to the responses. The results are presented in Table 6.2.

This important study shows clearly that both in the case of office and industrial buildings the largest problem is of inefficient layout. This lack of adaptability is generally not curable by rehabilitation and quickly leads to abandonment or, in the right conditions, demolition and redevelopment. There is no set of data to suggest how many buildings reach the end of their useful lives because they lack adaptability. However, it is clear from what we do know about the large growth of buildings over time, the number of buildings that are in use other than those for which they were designed, and the general increasing misfit of buildings with their function over time, that those which are not capable of accommodating change are very likely to be terminated. Whether it is possible to account for all the above factors in the first instance (at the design stage) is a moot question which may well entail building so 'loose' as to impair efficiency even in the initial years of functional life.

Land Values

The value of the land on which a building is erected may have a significant effect on the market price of that building. The value of land, therefore, and the factors which effect it are of great importance both to the design team and the building owner.

Table 6.2 Ranking of problems of older buildings.

Overall rank	*% of aggregate points*	*Problems*
Offices		
1st	35	Inefficient layout
2nd	21	Poor quality of M & E services
3rd (=)	15	Unsightly external appearance
3rd (=)	15	High energy/maintenance costs
5th	12	Inadequate car parking
6th	2	Poor quality internal finishes
Industrial/warehouse		
1st	40	Inefficient layout
2nd	32	Inadequate car parking/loading
3rd	13	High energy/maintenance costs
4th	10	Poor quality of M & E services
5th	4	Unsightly external appearance
6th	1	Poor quality internal finishes

Source: *Salway* (1986: 16)

The purpose of this section is to examine the nature of the relationship between the value of the land on which a building is erected and the value of the building itself over its life. This cannot be undertaken realistically without being aware of the nature of land, the factors which effect the value of land *'per se'*, and the changing patterns of land values at the macro level.[4]

Some of the distinguishing characteristics of land are held in common with other exchangeable commodities, others are unique to land. Under conventional economic theory it is held that land is relatively fixed in supply.[5] In absolute terms land is considered irreplaceable for no one piece of land is the same as any other. In theory, land is also held to be indestructible although the newly acquired characteristics of land around smelting plants, nuclear power stations and the like, sometimes render this assumption less tenable.

An acceptance of the broad principles of the orthodox economic analysis of land shows the clear importance of land resource allocation and regulation in the direction of national economic policy. Marshall himself, writing in the early nineteenth century, recognised this:

> A far seeing statesman will feel a greater responsibility to future generations when legislating as to land than as to other

> forms of wealth; and that from the economic and the ethical point of view, land must everywhere and always be classed as a thing by itself.
>
> (Marshall 1961)

If we consider the value of land to be the amount of money which could be fetched for the freehold or leasehold interest in a plot of land, particular determining factors can be distinguished. Supply and demand is arguably the most important determinant. For example, in the UK during the property boom of the early seventies, all other contributors to value were discarded as speculators lined up for the privilege of paying grossly inflated prices for property and land on the basis of increased demand alone.

The influence of location depends on what use the land is to be put, for example, a residential development would be unlikely to be placed near a large airport if there was some other site available, whereas an industrial development may well benefit greatly from being in such a position. Permitted use is governed by legislative restraints in the form of planning controls, by-laws and listed buildings, in order to provide some form of continuity in the environment. Miscellaneous determinants include the following: is the land freehold or leasehold? If the latter, how much of the lease is outstanding? What is the physical nature of the land, sloping or flat, firm or weak soil? Finally, what is the availability of transport, communications and services?

Marshall considered that the income from a landed property asset can be the aggregate of the returns from the bare unimproved site and from the improvements and buildings on the site. This provides a basis for understanding the behaviour of the returns from a building over specific periods of its life.[6] By developing an analogy of building life from a geographers image of an urban area as an organic form, Wood [1972], was able to construct a hypothetical representation of the relationship between building values and returns and those from the site.

Figure 6.1 is a hypothetical construct to illustrate the anticipated level of returns from a freehold building measured in terms of values at the very beginning of the building life. The additional returns vary throughout the various stages of the life of the building. The interest rate parameters w, x, y and z give the expected rates of real change in the levels of the successive returns to the buildings. The rate p is the expected rate of real value change in the plot value, the profile of which is the lower line

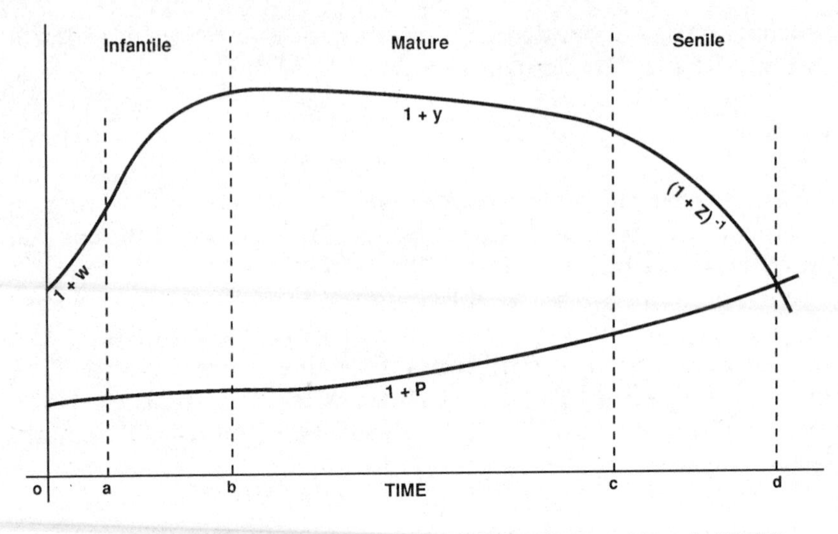

Fig. 6.1 Building value *vs* land value (*Source*: Wood (1972))

from *o* to *d*. In practice the phases, infantile, mature and senile are not always separately distinguishable nor are they always all present. In practice, the levels of returns are a function of all the land value determinants and are not always so tightly bound to the physical age of the building. The economic life of the building here is the time *o–d*, that period over which the net economic returns from the building are greater than or equal to those from the site. The elapsed time *d* is the optimum point at which the building should be demolished, it is the point at which the land and building profile intersects the site-value profile. If demolition occurs before *d* there is a loss in building value. If demolition occurs at some point after *d* there is a loss of site value as the full potential return from the site will not have been realised during that extra period. Either way, an economic loss is incurred.

It is clear from the diagram that the increase in site value can make the building obsolete. The owner would, other things being equal, choose to end the life of the building. Furthermore, Wood (1972) was careful to point out that it may be possible for a site value to increase at such a rate for it to intersect the building value before the senile period starts. The everyday examples which spring immediately to mind are those in which, in terms of Wood's construct, the land profile has an increased slope and intersects the building profile before the senile period sets in. These are to be seen in the central business districts of most growing towns and cities. Undeniably then, the increase in site value can have a terminating effect on building life, particularly in the central business districts.

Ideally, what is required in order to study this phenomenon is a set of data through which one could compare the value of land with the value of the buildings on it. At present there would appear to be no way of doing this which would come up with a more accurate assessment than the contemporary practice of making an individual calculation for each plot with its buildings as and when decisions regarding the future of the building have to be made. Without a statistical base and a dynamic model reflecting the actual change in values then, when these *ad hoc* calculations are carried out, at moments when decisions regarding the future of the building have to be taken, it will continue to be found in some cases that the decision to terminate the life of the building has been left too late, thus incurring a net economic loss. The absence of this data and, indeed, the impossibility of having it at all, means that the increase in site values will necessarily remain as one of the factors to be taken account of by owner, valuer, and design team on the basis of experience, theory and calculation applied to one moment in time.

Other Factors

The life of a building is determined by its relative value at the end point. The more obvious factors influencing this value have been discussed above. Value, as we have seen, is a function of marginal utility and supply and demand, all of which, in the market for buildings, are influenced by a complex matrix of underlying factors which are worthy of examination. It should be remembered that even though they are discussed individually for analytical purposes, in practice, they all operate together with varying degrees of interdependency.

Demographic Changes

That any increase or decrease in population has an effect on property values as such, is well known. The life of buildings, however, is affected not so much by absolute increase or decrease[7] in the national sense but more by migratory changes. A migratory shift will cause increased demand in one area and decreased demand in another. In practice, local and national migration operate together but there are three major migration patterns in the UK in the mid-twentieth century. Between 1961 and 1981 major conurbations decreased in population by 12.0%, sub-urban areas increased by 12.2% and rural areas increase by 26.6%. The first is the most important as it involves more people than the other two and because it has been a movement of people with no comparable movement of employment. It is one of the causes of traffic congestion. It has no direct effect on the regional population totals. However its indirect effects may be considerable.

Communications

Proximity to good means of communication is an advantage for most buildings. Transport patterns in the developed countries have radically altered in the past 40 years. Many countries have changed from being primarily dependent upon railways to being primarily dependent upon motorways and the road network. Industrial building, factories and warehouses have developed adjoining the road network. If the increasing road congestion or the energy supply situation causes us to revert to the use of railways or increase our reliance on air freight then many of the industrial buildings developed in the last five decades could become obsolete merely because of an increase in road transport costs relative to other methods. This may seem unlikely now, but 40 years ago the decline of railways seemed unlikely.

Planning Controls and Legislation

In many developed countries the planning profession exerts a strong influence over land use and value. A planning committee decision giving permission for the development of an industrial or residential estate on land that was previously used for agriculture will vastly increase the value of that site. Similarly in central business districts, permission for change of use of a small corner shop to offices will almost certainly result in demolition of the older building and redevelopment, regardless of its age, if profits are to be gained. Planning decisions and even discussions concerning proposed routes for ring roads have tangible effects on the value of property, fortunate or unfortunate enough to be in a particular place. The positioning of something like a ring road is governed by a very complex set of variables and the age of the buildings along the proposed route will not normally be top of the list.

Conclusion

Buildings and other constructed facilities are long lasting capital assets irrevocably bound to the land on which they lie. These two key factors lead onto a number of distinctive features of built environment obsolescence. It is highly likely that most buildings will become obsolete well before the end of their physical lives. The primary cause of building obsolescence are technological change and change in land values. Both of these are highly influenced by social and demographic change and by changes in transport and communications. In conclusion, these factors contributing to obsolescence are not operating individually but interacting in the context of the particular social, economic and financial climate of the time. Each firm or individual makes its decisions according to its own particular criteria.

Notes

[1] This question is considered in detail by Lynch (1972) in the context of his discussion of the relationship between time, the individual and the environment. See in particular, pp. 215–23.
[2] A seminal paper on the study of building life was that of Switzer (1963).
[3] Depreciation in leases and the role of the landlord tenant relationship, although of prime importance are outside the field of the book. It would be difficult to improve on the treatment of those areas given in Salway (1986: 113–127).
[4] Limitation on space does not allow a proper treatment here of the nature of land. An excellent overview of the subject may be had in Balchin *et al.* (1988).
[5] The amount that can be gained by reclamation or lost by flooding, erosion etc., is relatively minute – although there are some exceptions, such as Hong Kong, the Dutch Polders and Guyana.
[6] This point was developed in great detail in a Doctoral thesis by Wood (1972).
[7] Housing is more sensitive in this context; a sudden increase

in the section of population needing housing could well lead to the extension of the lives of many older dwellings through rehabilitation while the market for new housing catches up.

Part III

Supply-side Issues

Chapter 7

Behaviour of the Firm in Theory

Introduction

At the beginning of the 1990s there were about 180 000 construction 'firms' operating in the UK. In total there were about 370 000 registered companies operating in the economy, in a wider sense, in fields including manufacturing, services and retailing. Each of these is an economic unit with its own objectives, its own plans, its own possibilities. The central purpose of this chapter is to establish a conceptual framework which will help us to understand the behaviour of these economic units. Having, in a previous chapter, briefly examined the determinants of demand we now turn to the supply side of the industry, although much of what we have to say will apply also to the clients of the industry, from whom the demand for construction arises.

We will proceed by first considering the firm as an economic unit with an existence distinct from the people who work for it. Second, we will present a framework for the decisions to be taken by the firm. Third, we will review the theory of the firm insofar as it sheds light on the objectives of firms. Finally, we will consider some implications for the procurement of both new and rehabilitated buildings which derive from the theory of the firm.

The Firm as an Economic Unit

Firms start-up, grow, merge with other firms, and sometimes die. The firm's employees, and sometimes its owners, may change many times over, yet the firm as an entity may continue to exist. The stated objectives of a firm may not be congruent with the objectives of all or any of its employees. For our purposes here the firm need not be a corporation which manufactures a product. It can be any private or public sector economic unit which assembles inputs of *one* kind or another and produces some product or service for a client/customer. [*Note*: In the literature of economics the theory of the firm is concerned solely with the private sector, we have for our purposes expanded this to include public sector 'firms' who may be clients of the industry.] In this sense, the firm could be anything from, say, a soft drinks manufacturer to a social services department.

With the notable exception of private sector housing, the vast majority of construction work is ordered not by individuals but by 'firms' of one kind or another. Similarly, with the exception of the tiny proportion of work which is carried out by self-build or cooperative builder-user organisations, a larger proportion of all construction is undertaken not by individuals but by organised economic units, firms. Thus, it is important that we have some understanding of how firms operate, how they make their decisions, how they react to change in the economy, when they are ordering buildings or building them or designing them or estimating their cost.

A Framework for the Firm's Decisions

As we have already noted, this discussion encompasses a wide range of firms and types of organisation. As Morris (1985: 53) has described, in terms of size they range from one person private companies to multi-national firms with very large turnovers. In terms of legal constitution they range from private companies to partnerships to plc's (Public Limited Companies). There are also large differences in their decision making processes, their objectives, their product or service, their demand conditions and the type of competition they face. At an abstract level however, they do have some things in common. All need to organise some inputs of labour, materials, equipment and overheads to produce some good or service. All need, from time to time, to have access to finance. Most have to determine a selling price for their goods or services. All need to make some decisions about their objectives and to set about achieving these. It is with these sorts of issues that we are here concerned.

Taking all this into account Morris (1985: 54) produced a schematic representation of the 'typical' firm and the framework for its decisions. This is shown in Fig. 7.1. The diagram highlights both the firm's internal operations and the way in which it is influenced by external economic conditions. It is worth emphasising the circularity in the relationships shown. For example, the demand conditions affect the price, which affects the revenue (income), which affects the profit, which affects the funds available for investment, which affects (among other things) the amount available for research, which affects the costs of supply, which affects the profit . . . and so on! There are some weaknesses in using this diagram for our purposes. Namely, it does not deal with the setting of objectives for firms and it is intended for private sector firms and is therefore not really suitable for the social services department we referred to above. Further, it would be useful to identify those items which are capable of alteration by decisions taken in the firm, for example, physical and R and D investment.

Theory of the Firm

The theory of the firm has been the subject of considerable and unresolved debate for about fifty years. One major cause of this has been a lack of common purpose amongst those working in the field. The main thrust of the work up to the early 1960s was concerned with predicting and explaining changes in observed prices which result from changes in market and other economic conditions. In effect, the theory of the firm was not really a theory of 'the firm' but rather a theory, in which the actions of firms were prominent, of how prices and output volumes were related to market conditions. In the period since the 1930s this had become an increasingly abstract undertaking involving increasingly unrealistic assumptions with consequently diminishing application to the real world.

In defence of the traditional approach eminent economists such as Milton Friedman argued that it didn't matter that the assumptions were unrealistic if the predictions were correct. In other words we don't need causality if we can establish correlation. Most scientists would be unhappy with this kind of model and would regard it as only a temporary tool on the way to designing a model which would accurately *explain* the behaviour of the firm. However, the debate is only of peripheral interest to us as we are concerned with developing an understanding of the behaviour of the firms *per se*. To that end, we will focus only on the underlying assumption of the traditional theory namely, that the objective of the firm is to maximise its profits.[1]

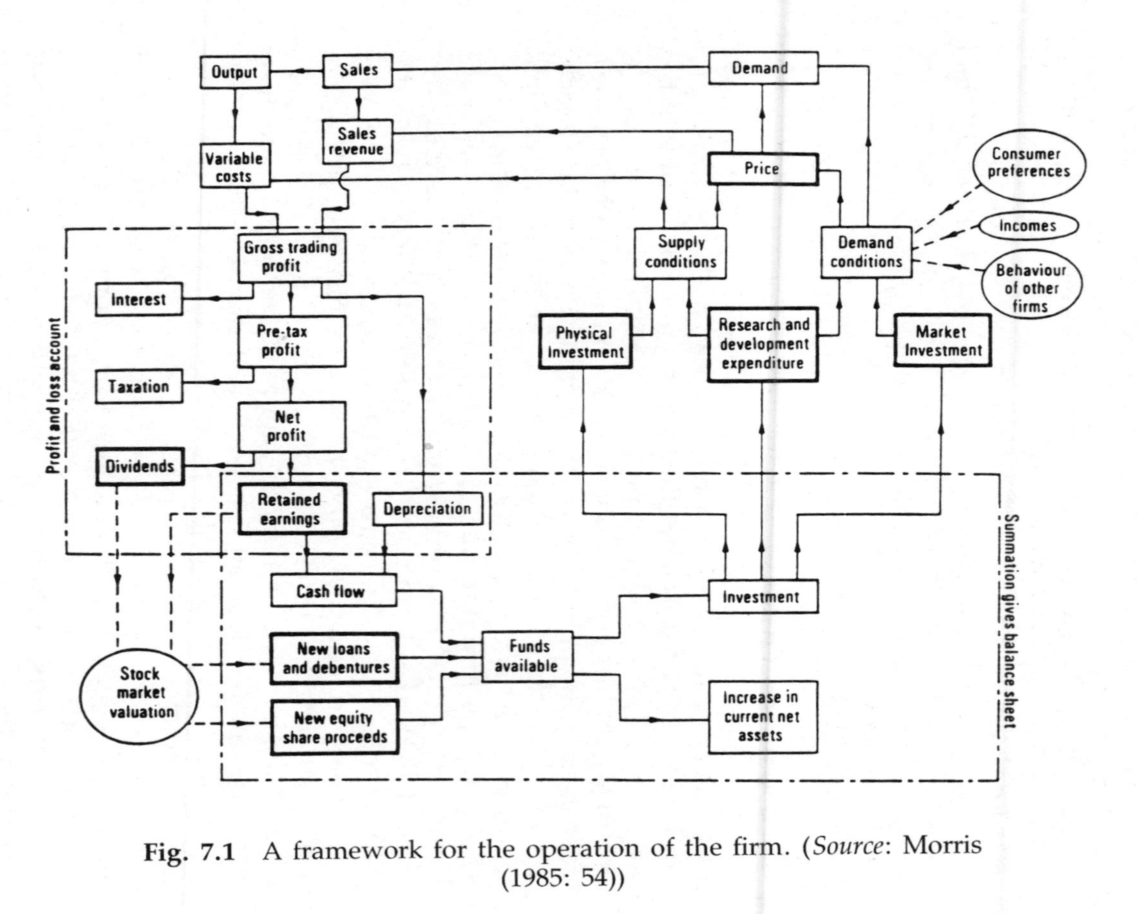

Fig. 7.1 A framework for the operation of the firm. (*Source*: Morris (1985: 54))

The most fundamental question for us to consider, is the extent to which profit maximisation is an accurate description of the objective of the firm. The assumption of profit maximisation was not based on empirical observation or evidence, rather it grew from the principle that if, in the long run in a perfect market, the firm did not maximise its profit, it would not survive. But of course, the perfect market is a theoretical construct and in reality many firms who do not maximise their profits do survive perfectly happily. Profit maximisation assumes the existence of a rational, utility-maximising owner-controller for the firm.

The criticisms of the profit maximisation motive focused on two particular problems. Firstly, the *ability* of the firm to maximise its profit and secondly, its *desire* to do so. The question of the firm's ability to maximise profits relates to the availability of perfect information on market conditions and ability to make certain forecasts in a risky and uncertain world. Clearly, in most cases neither of these are possible and thus it is highly unlikely that the firm is able to maximise its profit.

The question of the firm's desire to maximise its profit led to the development of the so-called managerial and behavioural theories of the firm. For our purposes we shall not distinguish between the two, we shall examine their collective contribution to our understanding of the behaviour of firms. The traditional profit maximising view of the firm assumed that it was controlled by a single entrepreneur in his or her own interests. Since the late 1950s we have become increasingly aware of the separation of ownership and control of firms, particularly large firms. The empirical evidence of this separation is not clear.[2] When firms are listed on the stock market their shareholders do ultimately control them by virtue of their power to vote out the directors. However, this power is rarely exercised and so it has been argued that in these cases ownership is effectively separated from control of the firm. In the construction industry about 90% of firms employ less than 25 people and it is considerably more likely that there is a lack of separation here.

The critics of the traditional theory assumed this separation to exist, and considered first, whether there would be any divergence of interest between the managers and the owners and second, whether conditions would allow the management to exercise sufficient discretion to pursue objectives not fully compatible with those of the owners.[3] This led directly to the suggestion that the manager-controllers would maximise other variables such as, sales revenue or growth of capital assets, rather than profits. Clearly, some minimum profit level would have to be achieved for the managers to be left alone long enough to do this. This notion was further developed to include the idea that the managers would in effect pursue their own utility. [4] This implies that they would be concerned with enhancing their own status, through their salaries, their personal offices and other accoutrements such as expense accounts and company cars. Again, this would need to be subject to some minimum satisfying profit level.

The behavioural theory has been developed further, in the context of large firms, to encompass the idea that the firm consists of groups or 'coalitions' of individuals who may have widely differing objectives. These groups negotiate with each other and with their superiors on the objectives to be pursued. Thus, a sales manager may be interested in maximising sales even if production is exceeding the capacity of the plant and the extra shift work needed is reducing the profit on the additional units sold. The firm's ultimate objective then depends on the outcome of the 'negotiations' among the various interest groups within it.[5]

The absence of the separation of ownership and control does not invalidate the criticisms of the profit motive. Recent empirical evidence on the threats of take-overs tends to support the managerial and behavioural theories where the profit motive becomes secondary to maintaining growth or ensuring that shares are not under-valued in order to ensure survival or avoid take-over (Nyman and Silbertson 1978).

Implications for the Construction Industry

In order to consider the implications of the above for the construction industry, we must first assess the extent to which the notion of 'the firm' as embodied in the literature has many features in common with the firm as encountered in the industry itself. Secondly, the behavioural and managerial theories of the firm indicate that profit maximisation may frequently be a secondary objective of firms. Is this the case in the construction industry?

Over ninety per cent of the construction firms operating in the UK would be classified as 'small firms'. Further, this structure is common to many countries. In firms of this nature it would seem much less likely that 'the firm' would have an existence entirely separate of those who own or work for it. Thus, it is more likely that the objectives of the firm would coincide with the objectives of its owners or employees.

Further evidence of the dichotomy between the firm in fact and the firm in theory has been presented in an important study by Eccles (1981). Eccles puts forward the notion of a 'quasi-firm' in the construction industry. The quasi-firm is a loosely organised set of sub-contractors who work from time to time for a main contractor. Thus the (relatively small) main contractor is empowered to tackle larger projects than would otherwise be possible. The main contractor chooses to work in this way rather than by expanding its own organisation (with consequent increase in fixed costs) up to the necessary capacity. This study demonstrated that the relationship between the main contractor and the subordinate elements of the quasi-firm (the sub-contractors) was usually long-term and rarely based on price competition. In effect, each main contractor had a panel of sub-contractors with whom it liked to work. These sub-contractors were used on each relevant project by the main contractor. More often than not they did not have to bid against other sub-contractors to win the work. Every few years, the main contractor would 'test the market' by holding a tender competition amongst the rival specialist sub-contractors to confirm the economy of the favoured sub-contractor(s) in a particular specialist area. This type of organisation is a termed a quasi-firm.

Although his work was carried out in the USA it does, on the face of it, seem resonant with anecdotal evidence from the UK industry. Thus, it seems that the contracting organisation constructing a project will frequently; not be of a size commensurate with the project, and not be a unified organisation but rather a temporary group of sub-contractors. Thus, we can conclude that, in the contracting part of the industry, the predominant type of firm has little in common with the theoretical large firm characterised by a high degree of separation of ownership and control which forms the basis of the literature on the theory of the firm.

Now, if Eccles is right and the price competition takes place only rarely, then this supports the idea that profit is a secondary objective on the relatively rare occasions when there is a tender competition. The firms' objective is to win that project which is necessary and sufficient to gain a place on the main contractors list. That is to say, to achieve market share. The sub-contractor can then look forward to a period of perhaps two or three years working for this contractor on a number of projects. It is during this latter period that, having increased his or her market share, the sub-contractor will seek to maximise profit. Similarly, a main contractor may sublimate the desire to maximise profit when bidding for a specific project, which, if won, will achieve penetration of a new market or increase the share in an existing market. The project being won however, and the main objective being thus achieved, the firm will then turn its attention to maximising profit within the constraint of the project conditions and the legal contract.

Thus, in conclusion, we may say that maximisation of profit, if it is a primary objective, is so only in the longer term. In the shorter term there is a very long agenda indeed, of strategic manoeuvres to penetrate new markets, to increase market share in existing markets, to put price pressure on competitors, to differentiate from competitors on grounds other than price. A detailed study based on interviews with 20 large UK contractors has been reported in Hillebrandt and Cannon (1988, 1989).

Conclusion

In sum, if we focus on the supply side of the industry in the short term we can see that other objectives such as, growth, securing a future workload, or moving into a new sector of the market, may prevail when firms are attempting to win any one particular contract. It is probably safe to say that having won a project the firm will be concerned to maximise the difference between income and expenditure. The importance of objectives other than profit may have curiously negative consequences in that the firm may be willing to strive for the other objectives at the cost of making a short-term loss. Carrying short-term losses is a high risk strategy. There is, in one sense, a contradiction at the interface between the industry and its client. The owners of the contracting firms have long term objectives related to the growth, survival and profit making capacity of the firm. The client of the industry, on the other hand, discounting the client's own firm-related objectives, has only short term objectives related to the delivery of the project(s) in hand.

Notes

[1] The seminal work of the time was Berle and Means (1932).

Their study entitled *The Modern Corporation and Private Property* is accepted as being the first to consider the managerial evolution. As Marris (1964) points out, there never was a managerial revolution. Instead, echoing the industrial revolution, there was a slow development from traditional, entrepreneur owning and controlling capitalism as it was replaced by a new form of economic organisation. With the formation of 'joint stock' companies day to day control of the firm moved away from the entrepreneur/owner/controller to a class of professional managers who were not generally owners of the firm. It was not until twenty years or so later that theories of the firm were developed which took this change into account (Downie 1958, Baumol 1959).

[2] A detailed study of the empirical evidence relating to the separation of ownership and control in UK industry has been published by Nyman and Silbertson (1978). This study showed that the extent of management control had been overestimated and based on too narrow a definition of ownership control. They also found evidence which indicated that owner controlled companies were in general more profitable than management controlled companies. They found that in the long run the owners do have significant control. However, when the owner is a financial institution, then if profits are maximised, the profits which are maximised will probably be those of the financial institution itself not those of industry in general. Their findings on owner control did not invalidate, and in fact lent some support to, the behavioural/managerial theories regarding the firm's objectives. The threat of take-over is a strong force pushing firms towards prioritising growth as their central objective. Thus even though they found less managerial control than expected they still did not find empirical evidence supporting the traditional theory.

[3] The most accessible primary treatment of the divergence of interests between owners and non-owning managers is probably Marris (1964) *The Economic Theory of 'Managerial' Capitalism*. Marris pointed out that there had heretofore been.

> 'not one comprehensive or rigorous study of the basic motivational forces determining business decisions in general: a fair literature examines the way the office affects

the home, little or none the way home and office life affect office decisions.'

(Marris 1964: 46)

Starting from the position that a manager is 'a different type of person from an entrepreneur, with different ideals and different personal values' (Marris, 1964: 6), he went on to examine the 'Motives and Morals' of managers. He did this from three perspectives, the psychological, the sociological and the economic. He proceeded to propose a model focused on two managerial utility dimensions, growth and security, in a context not of maximising or optimising but of 'satisficing'. The utility dimension of growth was represented by the growth rate of gross assets and that of security by the ratio of the market value of the firm to the book value of its assets (the market valuation ratio). In effect, this was a model for maximisation of growth rate subject to a minimum security constraint.

[4] Management discretion was the subject of a detailed study by Williamson (1967), *The Economics of Discretionary Behaviour: Managerial Objectives in a Theory of the Firm*. Williamson's model had application to public sector organisations.

[5] These points were developed in detail in the seminal work by Cyert and March (1963), '*A Behavioral Theory of the Firm*'. Using ideas from the then emerging field of organisation theory Cyert and March developed a general theory of economic decision making by firms, the basic concepts of which are well summarised in their chapter 6 pp., 114–27.

Chapter 8

Operation of the Firm

Introduction

This chapter is not about construction management, it is about the operation of the construction firm. Thus, it will include aspects of the management of construction projects but these will be a subset of the firm's operations. We are not here concerned with presenting quantitative techniques which may have application in planning or controlling resources. These are well presented elsewhere (Pilcher 1976 and Cormican 1985 for example). We are concerned with developing a qualitative understanding of how firms set about achieving their objectives and how they are constrained in this endeavour by their supply conditions and by the nature of the industry they operate in.

Accordingly, we will first recapitulate briefly on the objectives of firms and consider the most recent empirical and theoretical work on the economics of the firm. Second, we will describe how firms generate the revenue they need to survive and grow and generally meet their obligations to shareholders. Third, we will consider the control of costs once the firm has obtained the work and the potential profit margin has, in a stable environment, been set. Fourth, we will consider approaches to the measurement of the efficiency of the firm in terms of the concept of productivity. Fifth, we will introduce some basic ideas from 'portfolio management' to illustrate the problem of bidding for projects with uncertain outcomes. Finally, we will consider the strategic management of the firm in a dynamic environment.

The Economics of the Firm

Construction firms plan and carry out their operations in order to achieve their objectives. As we have seen, maximisation of profit may frequently be, in the short and medium term, secondary to other objectives such as survival, growth or just doing 'interesting' projects. In considering the operation of the firm we need to delve a little more deeply into the term 'maximisation of profit'. More precisely than maximisation of *profit*, the firm is interested in maximisation of *profitability* or return on investment (ROI). In construction firms it is difficult to measure the contribution to profitability of any individual project. This is due to the difficulty of apportioning the firm's fixed costs (overheads) to each individual project. Such apportionment can, at best, be notional. This has the further implication that the feedback on profitability of particular projects is fuzzy and as a result, difficult to use as a basis for improving the management of future projects. At the whole firm level, measurement of profitability presents fewer difficulties.[1]

Maximisation of profit is itself only possible in the presence of perfect information. The concrete, pursued objectives of firms are rarely so clearly defined as, nor are they necessarily congruent with, the formally stated objectives. All firms which employ more than one individual are organisations of human beings. The members of the firm interact, formally and informally, in cooperation and sometimes in conflict. They negotiate with each other in the setting and pursuit of the firm's objectives. There are many differences among firms including the type of work they do, their decision making processes, their objectives and the nature of the markets they operate in. However, all firms organise personnel, equipment, materials, entrepreneurial skill and finance to produce some sort of good or service. Thus, there will be a technical dimension to the work of the firm in the sense of planning, organising and constructing buildings or parts of buildings. There will also be non-technical dimensions to the firm's work in the sense of arranging finance for its activities, marketing, and human resource management for example.

It is useful to consider both the *process* of operating the firm and the *environment* in which it operates (Clarke and McGuiness, 1985). The environment in which firms operate is, we will demonstrate, characterised by high levels of uncertainty, very imperfect information and by frequent instances of what is called 'small numbers bargaining'. This situation is a long way removed from the base-line assumptions of large numbers of actors operating with near perfect information in a frictionless economic world.

We have, in the previous chapter, touched briefly on the managerial and behavioural theories of the firm. Current research in this field has moved away from considerations of the firm as a 'black box' (with resource inputs and profit outputs), to an approach which attempts to analyse the forces and motivations operating within the firm which result in its observed behaviour. According to Clarke and McGuiness (1985: 8) this work appears to fall into three areas.

First, the analysis of principals and agents. The so-called principal-agent model is used to describe the situation where one party to a contract (known as the 'principal') engages another party (the 'agent') to carry out work or act on his or her behalf, in a situation where there is unequal ('asymmetric') information (Arrow 1984, Jensen 1983, Holmstrom 1979, Shavell 1979). Clearly, this is of central importance in the construction industry with its heavy reliance on subcontracting and given the empirical evidence of the existence of the, so-called, 'quasi-firm'. The principal-agent model takes explicit account of uncertainty and information. Principal-agency relationships are not restricted to subcontracting, they exist between, among others, shareholders and managers, employer and employee, client and advisor. The model assumes that both parties act in their own self interest. The principal delegates to the agent the responsibility for selecting and implementing a course of action. The principal compensates the agent and lays claim to the residual outcome of the agent's act. Given that the objectives of the two actors are not in complete harmony, the principal's problem is to design a contract which minimises divergence between the two, taking into account the degree of risk aversion of both parties.

Second, the study of 'firms versus markets' (Williamson 1975, 1981, 1984). The thesis of this field is that the reason for the existence of a firm is that it presents a method of producing the good which has lower transaction costs than would be incurred if the same good were to be assembled by the buyer in the market. For example, bicycle firms exist because they can, through bulk buying and other economies of scale, manufacture and assemble a bicycle more cheaply than I can buy the components in the market and assemble them myself. This explains why it is worth having a 'management' to organise the bicycle firm. The theory goes on to explain management hierarchies in terms of the gains to be made by allowing people to specialise in their areas of comparative advantage. The importance of this work is that it focuses on the economising of resources used in 'managing' (planning and controlling) as opposed to 'producing' the good in question.[2]

Third, is the theme of strategic firm behaviour. 'Strategic' here has a special meaning. It refers to an action taken by a firm with the specific intention of influencing other rival firms behaviour. Thus, by definition, the study of strategic firm behaviour is in the realm of oligopolistic interdependence, small numbers bidding and bargaining.

Generation of Revenue in the Short-term: Sealed Bidding

The firm generates revenue by selling its product. Contractors generate revenue by 'winning' projects. In the developed world the majority of construction projects are awarded to contractors as a result of some sort of tendering process. The tenders may sometimes be in open competition or, more usually, invited from a prequalified list.

Projects are not always awarded to the lowest bidder, primarily for three reasons. First, because there are occasions when price is not the most important factor in awarding the contract. Second, the lowest bid may be an error and the client's advisors may feel that the project could not be built for this price, thus, involving the client in the potential bankruptcy of the bidder in question. Third, there may be other political reasons for awarding the contract to a particular contractor. This sometimes occurs in international projects where there may be links with other diplomatic initiatives. In the latter case, a special relationship with one country or contractor may result in the lowest price anyway.[3] Thus, in the majority of cases, contractors win projects by bidding. More specifically, by submitting sealed bids. Hence, in practice most firms will not know the prices being asked by their competitors until after the project has been won and the price has been set. Hence, in the context of the theory of price determination, bidders will have no information about the prices being asked by their competitors.

Our interest, it should be restated, is in how firms generate revenue through the bidding process. Selling the goods, generating the revenue through bidding is not frictionless or cost free. Thus, we are concerned with the constraints, possibilities and uncertainties placed on the firm in its real world market situation, and with their implications for the operation and management of the firm. Clearly, the bidding process is both complex and important. We will approach it here by considering first the rules of the market and then by analysing the bidding situation itself through the medium of bidding models.

Rules of the Market

For the purposes of our argument let us focus only on those projects where the contractor is appointed by a conventional 'closed' bidding or 'selective tendering' system.[4] In this system only contractors from a prequalified list are invited to bid. In order to gain a place on the list, each contractor needs to convince the client's advisors that it has the technical and managerial competence to undertake projects of this scope. The competition to get on the list may be largely, but not totally, non-price. The client's advisors will take into account bank references, credit checks, quality, scale and 'success' of previous projects and any personal knowledge they have of the firm. When the client decides to appoint a contractor for a specific project, a small number (in the UK usually between 4 and 8) of firms are invited to tender for the project.[5] From this point on, it is assumed that the competition is purely in terms of price. It is assumed that there will be no collusion among the bidders, no cartel and no revealing of prices.

Amendments, conditions and the addition of so-called 'non-price features' to tenders are usually, in the UK at least, strongly discouraged. This includes tenders which, when opened, are found to include suggestions for different forms of construction or for a shorter time scale to complete the project. This discouragement is meant to render the tenders comparable so that the decision to appoint is a straightforward one. It must be said that it is also very stifling of innovation. Is this a fair price to pay for 'ease of comparability' of bids? Surely it is not beyond the capacity of trained professionals to compare a set of bids on the basis of more than one criterion? It is also assumed that if the firm does not wish to tender, that they will say so without prejudice to their chances of being invited to bid for subsequent projects. Depending on the size of the project, it is usual to allow between 3 and 6 weeks for the preparation of bids. These are general rules. The particular set of rules adopted depends on the procurement system used. Franks (1986) describes a variety of procurement systems.

Bidding Models

Let us focus now on the contractor. The range of objectives which a contractor might hold at a particular moment or with respect to a particular project, could include the following: First, to maximise profit on each individual project. Second, to maximise return on the capital invested in the firm. This would involve an optimisation exercise by attempting to win a lot of projects at a very low markup. Third, to minimise losses. Fourth, to maintain production and the employment of the firm's labour force at almost any cost. Fifth, to enhance the personal status of the owners/managers by undertaking prestige projects. Sixth, to gain a foothold with a new client who is the potential source of much future work. Seventh, to gain a foothold in a new geographical area. Eighth, not to 'offend' a valued client by refusing to bid even though the firm does not have the spare capacity to take on the project should its tender be accepted. On being invited to tender, the range of possible courses of action has been outlined by Skitmore (1989b) as follows:

(1) Return tender documents.
(2) Enter a 'cover' price [not strictly within 'the rules'].
(3) Calculate a rough cost estimate and add a markup.
(4) Calculate a detailed cost estimate and add a markup.
(5) Add 'non-price' features.
(6) No bid.

On being invited to bid, the first step is for the contractor's senior management to review the project in order to determine whether the firm should go ahead to prepare a bid. This review will take into account, *inter alia*, the technical, financial and managerial capacity of the firm, its current workload, its forecast of future workload and the likelihood of being successful on other bids currently being prepared, the location and nature of the project in the context of the management and market strategy of the firm, the client and the availability of resources generally. Understandably, there has as yet been no successful attempt at building a decision support system to take all of these matters into account. However, a little of the complexity of the bidding situation may be revealed through the study of bidding models.

The formal study of bidding models in construction applications began with the seminal paper written by Friedman (1956). Skitmore (1988), a leading UK researcher in this area, has pointed out that since then the literature has grown so large that it is probably now beyond the reading capacity of new researchers entering the field. Fortunately, there are some useful reviews of the literature, for example, Englebrecht-Wiggans (1980), King and Mercer (1988), Raftery (1985) and Skitmore himself, (1988, 1989).

Eleven years after Friedman's original paper Gates (1967) published a paper which took issue with one of Friedman's assumptions and since then a vigorous debate has ensued in the pages of the Journal of the American Society of Civil Engineers. The subtleties of that debate are not germane to this chapter and further, they do not alter the fact that the general principles underlying bidding models as laid down by Friedman have not changed. Friedman and Gates' models were both designed to assess the probability of winning a job with a given bid and, in effect, to locate the optimum bid. That is, the bid which maximises the profit of the contractor over time. Thus, in conventional bidding models, we enter the field very late in the game, i.e., after the complex decisions described above have already been taken and the firm has, not only decided to bid but also, decided that it wants to win the project.

In aiming to win a project the contractor hopes to submit a price which is low enough to win the project but high enough to make a profit. The bid may be resolved into two components as follows:

$$\Sigma \text{ Costs} + \text{Mark-up} = \text{Bid}$$

Hence, in bid preparation the first task is to compile a net cost estimate. This should include intangible and indirect costs such as overheads, finance charges etc., as well as the more obvious direct costs of personnel, equipment and materials. The net cost estimate will have associated with it, a margin of error. Notwithstanding some claims to the contrary, there is, to best of this writer's knowledge, no reliable empirical estimate of the margin of error surrounding a contractors net cost estimate. If, for given projects, a firm consistently produces a higher cost estimate than most of its competitors then, in the long run, it will go out of business. In the long run, firms should tend towards equal technical efficiency. Hence, in compiling bidding models it has been taken as acceptable to assume that, for a given project, the net cost estimate of each competing firm should be the same. Once this estimate has been compiled then the firm makes a tactical decision on the level of the mark up. Clearly, the higher the mark-up, the lower the chance of winning the project. Using some simplifying assumptions, for example, that at zero% mark-up the bidder is 100% likely to win, Park (1979) illustrated this point in the following way.

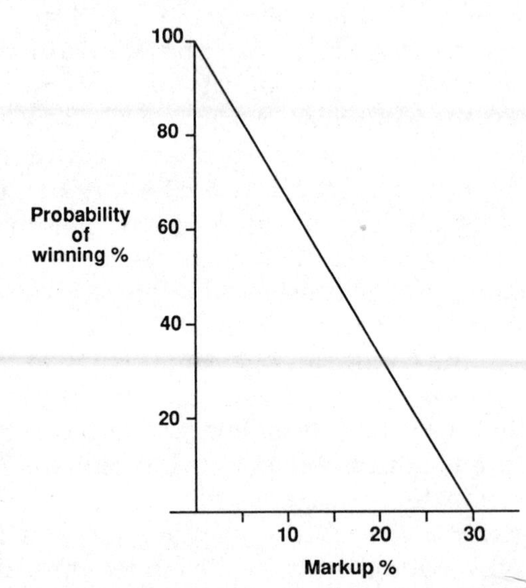

Fig. 8.1 Relationship between level of mark-up and likelihood of winning (After Park (1979))

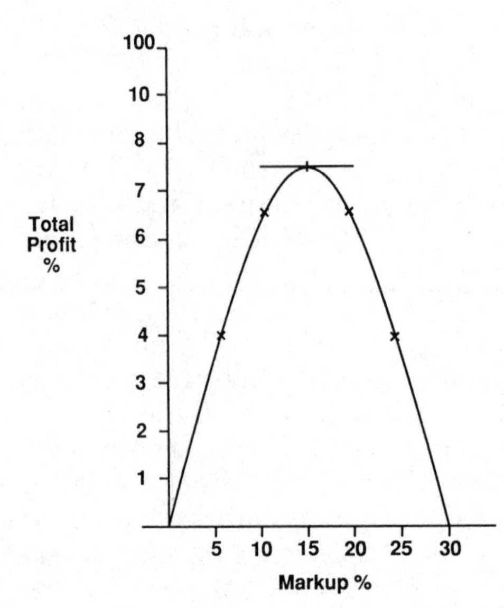

Fig. 8.2 Optimum mark-up against one competitor (After Park (1979))

Consider a firm bidding for a project against just one competitor. If we assume that at zero% mark-up the firm is 100% likely to win the project and that at 30% markup the probability of winning is zero, then the situation may be illustrated as in Fig. 8.1. Using graphical calculations we can see that at a mark-up of 5% there is an 83% chance of winning. Hence, the expected profit from this would be 83% of 5% or 4.5%. Similarly, at a mark-up of 10% the potential profit would be 66% of 10% or 6.6%. At a mark-up of 25% there would only be a 16% chance of winning, thus the profit would be 16% of 25% or 4%. These results are graphed in Fig. 8.2, from which we can see that, for these assumptions, the optimum mark-up is 15% and that this mark up maximises profit at 7.5% of turnover.

If we increase the number of competitors the probabilities of winning at intermediate mark-up levels are affected as shown in Fig. 8.3. Thus, given our initial assumptions and depending on the number of competitors, we can graphically optimise the mark-up to maximise profit levels. Bidding models are based on the principle that there is one net cost estimate for the project and that, for a given project, the only variation between bids of different firms is the level of mark-up. To build a bidding model the firm compiles a history of its recent bidding performance as shown in Table 8.1. In order to do this, it needs to know, not

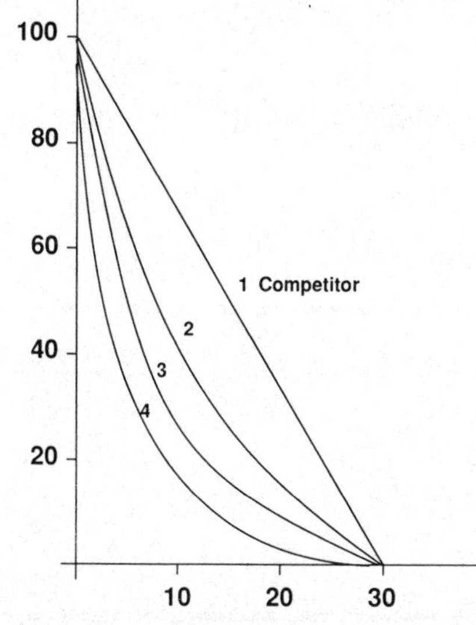

Fig. 8.3 Effect of number of competitors on probability of being the lowest bidder. (After Park (1979))

only who its competitors were for each project, but also, what their bids were. Assuming that the cost estimate is common, it is a simple step to calculate the mark up level of each competitor. If the data from Table 8.1 is compiled for many projects then the firm will probably find that it bids against some of its competitors repeatedly. Thus, it will be possible to plot the pattern of mark-up adopted by consistent competitors and to build a model which suggests a mark-up with optimal chances of beating known competitors for a given new project.

The disadvantages of bidding models are well-known and documented, (Raftery 1985). It is for example, extremely unlikely that at zero% mark up there is a 100% chance of winning the project. Given, that, in the long run 'we are all dead', and that all firms operate in dynamic short run situations, there is no reason to assume, that, at any given moment, firms will have equal technical efficiency and thus, equal net cost estimates. Add to this the possibility of errors, and it is certain that to achieve a 100% chance of winning the firm would need to use a negative markup, in other words, to bid at below its own (estimated) real cost. Bidding models are not dynamic. Data may be collected over, say, a two year period. Then the model is based on a two year 'snapshot' of bidding history. Market conditions and competitors' behaviour change more rapidly than this. Hence, the optimum mark up suggested by a bidding model in practice may well be an out-of-date optimum in a given contemporary situation. The problems of bidders will not be solved

Table 8.1 Data for bidding models: various authors

Project No.1

C_o = Contractor *O*
E_o = Net Cost Estimate
B_o = Bid

CONTRACTOR	*COST ESTIMATE*	*BID*	*MARK-UP*
C_o	E_o	B_o	$B_o - E_o\%$
C_1	E_o	B_1	$B_1 - E_o\%$
C_2	E_o	B_2	$B_2 - E_o\%$
—	—	—	—
—	—	—	—
—	—	—	—
C_n	E_o	B_n	$B_n - E_o\%$

$B_n - E_o\%$ = Bid/Cost Ratio

until there is a comprehensive decision support model which can capture the full complexity of the bidding situation.

Implications for the Operation of the Firm

Thus, sealed bidding places a number of constraints on the firm in its endeavours to generate revenue and 'sell' its product. First, firms have very little and very coarse information on the market, in particular, they have little or no knowledge of the prices being charged by their competitors. It is not safe to assume that the firm can sell all of its output. In the short run, many (inefficient) firms will be operating below capacity due to their not being able to win projects. It is very difficult, perhaps impossible, for the firm to plan and to 'smooth' its future output, due to the fact that it may be tendering for many projects simultaneously with little idea of how many of its bids are likely to be successful.

Control of Costs in the Short-run

Having won a project, the firm's task is one of controlling the cost of the project in order to maximise the difference between expenditure and income, within the constraints of specification and contract. Both expenditure and income are phased. The circumstances of most construction contracts are such that the client makes interim payments to the contractor for work in progress. The payment frequency is normally monthly, with a small percentage (of the order of 5%) being retained until the contract has been satisfactorily completed. Assuming 5% retention and monthly payments, then in theory, a firm has to have available to it the finance necessary to pay its short term debts such as wages and raw materials for one month until the next interim payment and to be able to pay for the 5% which is retained until the end of the contract. In Pilcher's (1985: 288) words, 'The need for working capital arises because of a delay between the expenditure on resources and the payment for goods subsequently provided'. Thus, the firm does not usually have to finance the entire project. Instead, it has to have available to it, sufficient finance to pay for its working capital.

The picture presented above is oversimplified as in fact the contracting firm will take advantage of 'free' trade credit from its suppliers. Under these arrangements the contractor may not have to pay for goods supplied for 30 or even 90 days. By careful manipulation of credit periods and interim payments the contractor may minimise the working capital necessary. The firm may even receive payment from its client before it needs to pay its own bills. In this way, small firms with access only to small amounts of working capital may take on projects which are very large indeed compared to the size of the firm. In effect, there is a large multiplier between the amount of working capital and the size of project which the firm may undertake. If we assume, perhaps unfairly, the availability of performance bonds and bank references and a retention rate of 5%, then this multiplier is of the order of 20. Hillebrandt (1984: 275–82) discusses the ratio between capital and turnover. From her figures for the late 1970s in the UK, it is possible to calculate that, for publicly quoted building and civil engineering contractors, working capital was about 15% of turnover.

In the case of a single project the working capital requirement may be calculated graphically as shown in Fig. 8.4, where the curve represents the cumulative costs incurred by the firm and the stepwise function represents the cumulative payment to the firm. The working capital requirement for a single project can be extracted in the form shown in Fig. 8.5. From this it is a simple step to compile the information for all of the firm's current and planned projects to produce a forecast of the firm's total need for working capital.

Productivity

In building economics there are few technical issues which can generate more muddled thinking than productivity. While there is general agreement that 'high productivity is a good thing', there is far less agreement on definitions and methods of measurement of productivity. Newspapers, and sometimes more authoritative sources, report on something called 'low productivity' when what they frequently mean is *low* labour productivity. This is often implied to connote 'inefficiency'. In fact, in a labour intensive process, 'low' labour productivity may well be an indicator of *high* efficiency in the use of resources.

In the following paragraphs we will first consider the concept of productivity. Second, we will examine means, ends and reasons why measures of productivity are useful. Third, we will describe

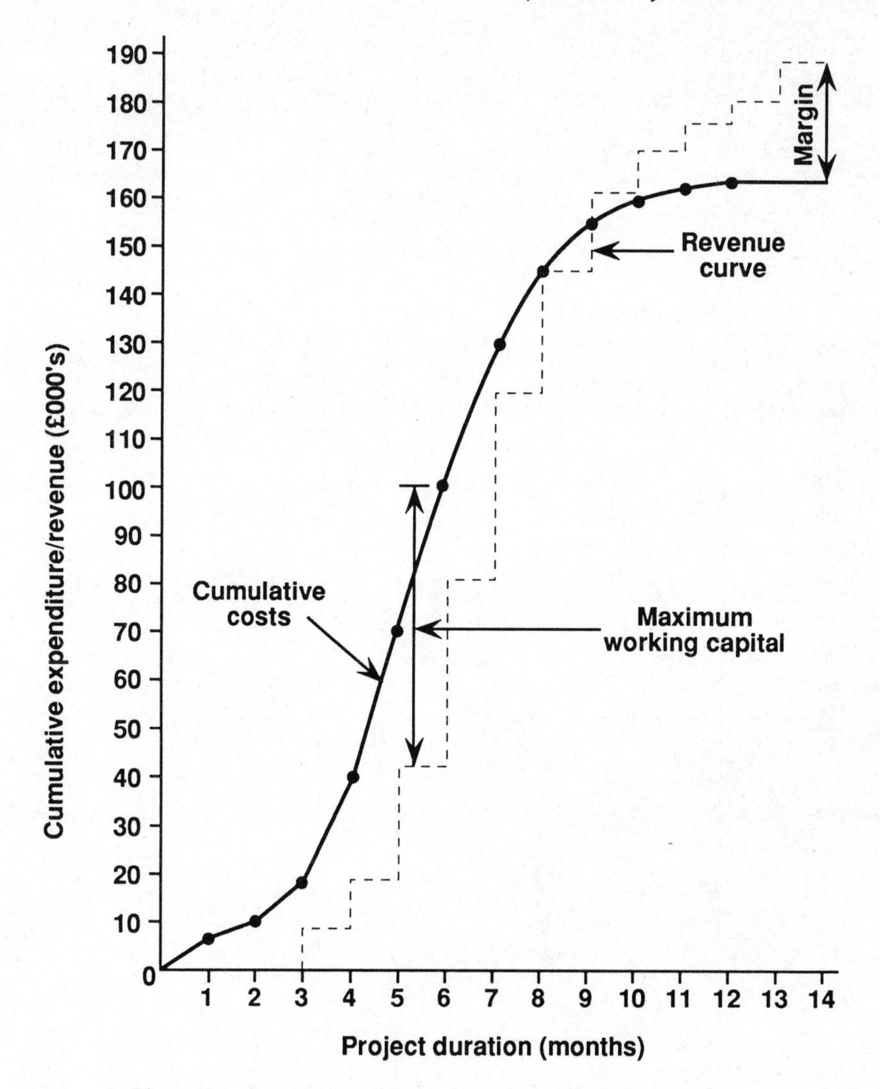

Fig. 8.4 Financial curves for a construction project. (*Source*: Pilcher (1985: 299))

the measurement of single and total factor productivity. Fourth, we will present some common misuses and misunderstandings of productivity. Fifth, we will describe some of the particular features of construction productivity.

What is Productivity

In general, productivity is a measurement of the relationship between the outputs of a production process, an industry or an

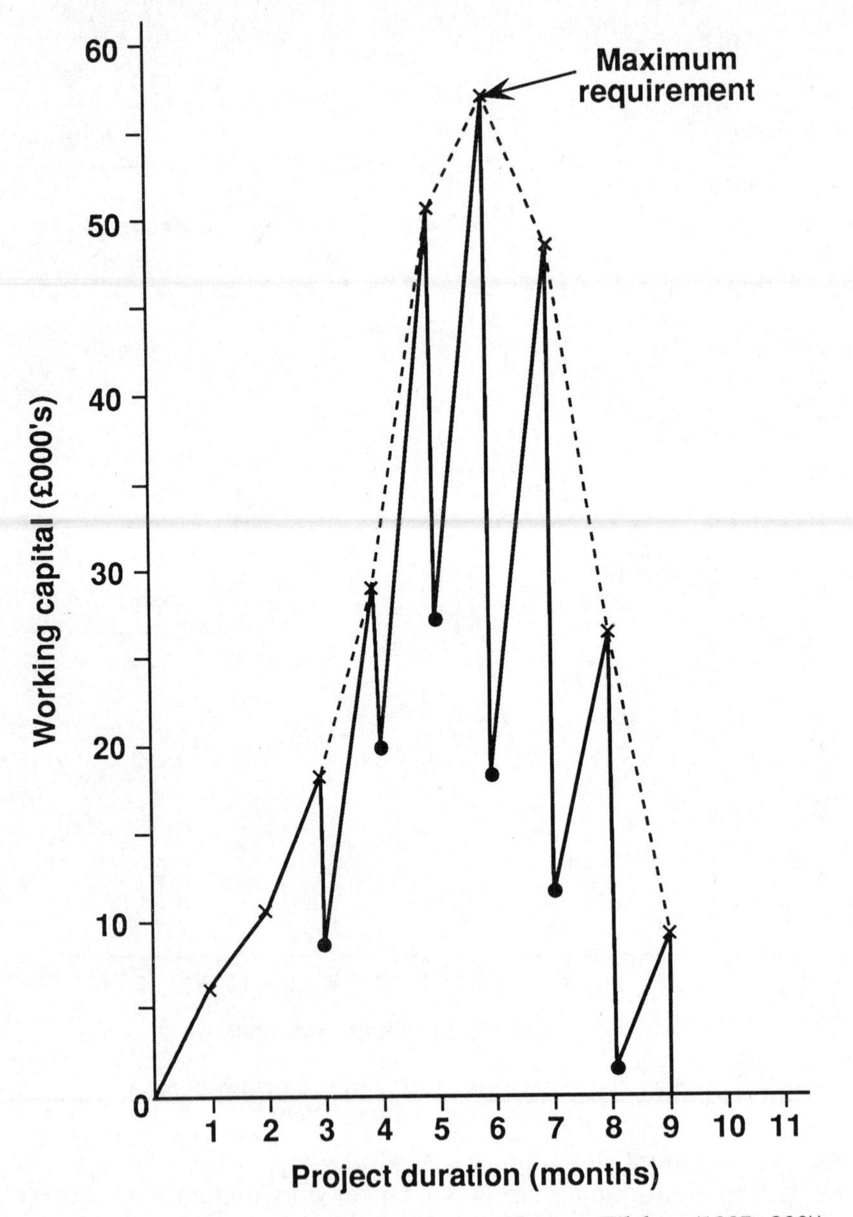

Fig. 8.5 Working capital requirements. (*Source*: Pilcher (1985: 300))

economy and the corresponding factors of production (inputs) which generate the output. Single factor measures relate the output to only one type of input, for example, the number of labour hours to produce a finished motor car, or the number of kilowatts of energy needed to produce one tonne of Portland cement. Both of these examples are also examples of so called 'physical partial' methods, that is, they relate the physical output (numbers of motor cars, tonnes of cement) to just one of the input resources (labour or energy).

The most common single factor measures are labour productivity and capital productivity. Total factor productivity is an attempt to account for all inputs (usually, labour and capital) by producing a weighted average.

There are two aspects to productivity, productive and allocative efficiency (Lowe 1987). Productive efficiency is concerned with minimising cost, for a given level of production, by optimising the costs and quantities of individual inputs. Thus, productive efficiency has application at the level of the project, the firm and the industry. Allocative efficiency is concerned with the optimal allocation of scarce resources to the various productive sectors of the economy. For example, are large amounts of labour best used in the construction industry? Could this labour make a better contribution to the economy by being used in another sector, say, with a higher 'value added' and consequent higher contribution to GDP *per capita*? Allocative efficiency is usually measured with respect to the *'pareto optimum'*, the condition whereby, at the level of the national economy, it is not possible to make anyone better off without making someone else worse off.[6] In consequence of this and the approach to building economics taken in Chapter 1, we shall here be concerned mainly with aspects of productive efficiency.

A Means to an End: Why Measure Productivity

Ratios, in this case between output and input, have no absolute meaning. They are relative measures and have meaning only in relation, either, to changes over time, or to other ratios at given points in time. Thus, productivity measures are useful for comparison. For example, changes in the number of labour hours needed on site to construct domestic dwellings or a unit area of built space may indicate a change in technology or organisation. Assume, for the purposes of argument, that the number of hours necessary goes down. This may indicate that labourers are working harder, or that new materials are easier to erect, or that more work in prefabrication is being done off-site, or indeed that contractors now use more equipment and maybe the labourers are working less hard and still building more space per hour. Why then measure productivity?

Four types of comparison may be identified. Inter-process comparison, in other words, compare the productive efficiency of, say, domestic dwelling unit type 'a' and type 'b'. Inter-firm comparisons are vital to the relative productive efficiencies, and thus the survival, of different firms. Inter-national comparisons help us to judge overseas competition. Inter-industry comparison helps us to assess the relative importance of the industries in the economy.

High productivity is a means not an end. Reducing the labour content of a process assumes that we have something better to do with the time. This might be the case. It also might be the case that the labour which we shed does not have anything else to do and becomes unemployed. This may ensure the survival of the firm/industry. The means to this survival has been the (short-and, possibly, long-term) loss of jobs.

A graphic example of this is provided by the performance of the UK steel industry in the 1980s. In the 1970s the UK steel industry was one of the most inefficient in the world and produced 40% less steel per worker than the West German industry, for example. At the end of the 1980s it was one of the most efficient producers in the world. Between 1980 and 1989 the labour content per tonne of steel in the UK industry fell from 14.5 person hours to less than 6. In the same period the workforce fell by 66%.[7] We can postulate that some of these workers found jobs in other industries, some, in spite of difficulties, may have moved to cities where they could find work, and some may have remained unemployed. During this period, unemployment rose and then in the late 1980s it began to fall again. It could be argued that an increase in labour, capital and total factor productivity in this period, caused short term job losses but that the increases in productive efficiency would lead to the economic strength needed to increase employment.

Measuring Productivity: Single and Total Factor Productivity

In order to measure productivity we need to measure the quantities of both the outputs and the inputs of the production process in question. The more homogenous the output the easier it is to measure. For example a figure for barrels of oil has more meaning than a figure for the number of buildings completed. Clearly, in the majority of cases where the output is not homogenous we will need to use a monetary value of the receipts for the output. This value will need to be adjusted for the changes in stock. In other words, output value will be the value of receipts plus the value of increases (minus decreases) in the stock of finished goods. Most builders do not carry a stock but this will be an issue for private housebuilders.

Construction presents particular difficulties due to the extent of subcontracted work. It is important, in measuring outputs, to exclude any receipts for subcontracted work in order to avoid double counting. For example, if contractor A's output measure includes £x received from contractor B for a sub contracted job. This £x will be counted again when contractor B records its output as the total of its receipts including £x which it happens to have subcontracted out (Weber and Lippiatt, 1983).

Separate indexing of output and input values is also important. Weber and Lippiatt (1983: 3) quote an example, suppose that receipts for output increase due to increased prices, all other things being equal, i.e. while the value of output and input quantities and input prices remain the same. Applying the input cost index to the output measure (the receipts) would leave them unchanged and thus indicate an increase in output. This would cause productivity to appear to increase while the real ratio between output and input would in fact remain the same.

Conversely, suppose there was a real productivity increase, i.e. output quantities increased while input quantities remained constant. If input prices rose slightly, but due to market pressures, output prices remained constant, then if an input cost index were applied to deflate the output prices the value of the real productivity increase would be diminished. Table 8.2 illustrates the calculation of single factor productivities for labour, materials and capital.

A similar double counting problem arises with the measurement of material inputs. In order to produce accurate measures of the relationships between real inputs and outputs we need to isolate the amount of 'value added' during the production process. Thus, the use of factory finished components in building could tend to push apparent productivity upwards as the labour inputs for these components would appear in the statistics of another industry, Hillebrandt (1984: 222).

The most widely used, and abused, single factor measure is labour productivity (Sugden 1978, Langford 1982). Average labour productivity is arrived at simply by dividing the quantity of output by the quantity of labour needed to produce it.

Table 8.2 Single Factor Productivity Measurement: An Illustration

Productivity Component	*Quantity* Period 1 (1)	Period 2 (2)	*Index of change* (3) = (2)/(1)	*Single Factor Productivity Measure* (4) = 1.100/(3)
Output	500	550	1.100	N/A
Input				
Labour	130	150	1.154	0.95
Materials	100	110	1.100	1.00
Capital	80	70	0.875	1.26

Source: *Weber and Lippiatt* (1983: 11)
Note: Quantities can be denominated either in physical units of measurement or in quantity index numbers derived by dividing current dollar values by their appropriate price indexes.

$$ALP = O/E$$

where ALP = average labour productivity.
O = output, measured in monetary units deflated by an output index if being used for comparisons over time.
E = the number of employees in the industry.

Clearly, this glosses over many complications of measurement such as, how to deal with labour only subcontractors who do not show up in the industry statistics, and how to take account of hours worked, as opposed to, hours paid for. Average labour productivity, although relatively easy to calculate, may be misleading in construction due to the small number of directly employed workers who are supplemented by subcontractors not appearing in the returns on inputs. Lowe (1987: 104) contends that even if the output is measured in value added terms the average labour productivity of construction work may tend towards infinity.

Capital productivity is the rate of return on the capital invested, given thus:

$$CP = P/C$$

where CP = capital productivity.
P = profit.
C = capital invested.

Table 8.3 Hypothetical Data on Output and Input Quantities, Input Prices and Cost Shares of an Industry for Two Time Periods

		1981			1982		
Symbol	*Variable*	*Quantity*	*Price ($/Unit)*	*Cost Share*	*Quantity*	*Price ($/Unit)*	*Cost Share*
Y	Output	100	—	—	110	—	—
X_1	Labour	40	15	0.115	30	16	0.082
X_2	Capital	30	20	0.115	40	22	0.150
X_3	Materials	2000	2	0.770	2200	2.05	0.768

Source: *Weber and Lippiatt* (1983: 16)

Clearly, for comparisons over time, it is important that some sort of discounted cash flow method is used. There are major difficulties in measuring, with any accuracy, the amount of capital invested in construction firms. Fixed capital is, in principle, easier to measure than working capital. The fixed capital requirements of construction firms though, are relatively low as we have seen above. The exception is firms dealing speculatively in land and property and in the private sector housing market where stocks of land and unsold buildings may be held.

Further difficulty arises from the subjective nature of fixed capital valuations and the amount of latitude used in calculating depreciation, for example. Many takeovers of firms take place just because the fixed capital assets of firms have been severely undervalued. For traditional construction firms working capital is relatively important but is not reported publicly and is difficult to measure even internally. All that said, for firms operating in capitalist economies capital productivity is very important. Put simply, if shareholders do not receive an adequate return on capital invested, none will be invested and the firm will not survive.

Total factor productivity is the weighted average of the inputs related to the outputs. This has the potential to produce an effective measure of productive efficiency. Theoretically straightforward, there are a number of practical difficulties with collecting the relevant data. In principle, we need to establish the input quantities, prices and cost share for each input to the production process as indicated in Table 8.3.

In researching construction industry productivity Lowe (1987) adopts a two factor production model using labour and capital inputs. Chau and Walker (1988) use a four factor model considering labour, materials, plant and overheads.

Limitations of Productivity Measures

There are fundamental problems with the use of single factor measures of productivity such as labour productivity. First, they do not deal well with the effects of factor substitution. Consider a simple production process which uses two inputs, labour and capital. Assume that more capital is invested in equipment, output increases and other things remain equal, that is output prices and labour input prices and quantities remain constant.

In this situation, labour productivity will increase when the increase in output has been caused by the new equipment. A second, and to an extent, related problem is that of change in relative prices.

Consider the same two factor production process. Assume that the price of capital equipment increases. The firm's managers could decide to maintain production at a given level by substituting labour for capital. In this example labour productivity will appear to fall when what has actually happened is that a change in relative prices of inputs has caused factor substitution to take place. Thus, in relation to labour productivity in particular we can conclude that it is a measure of the labour intensity of the production process and, in itself, implies nothing about the efficiency of the allocation of resources.

Total factor productivity (TFP), in theory, resolves these problems. In practice there are real difficulties in collecting data to produce accurate measures of TFP. Chau and Walker (1988) have presented some preliminary results for a TFP study of the Hong Kong industry using a four factor production model. Their paper illustrates well the difficulties of data collection. First we need to be able to identify the inputs separately. Second we need to attach weightings for the cost share of each input to the overall production process. In other words we need to know the proportion of total output accounted for by labour, materials and capital if they are the inputs we are studying. Third, we need to have separate indexing of input and output values as discussed earlier. We also need separate indexing of each input factor so that we can observe any changes in relative prices. Fourth, we need to have a measure of how the profit margin changes over time. Without this, we would not be able to tell if an increase in the value of output was merely due to an increase in profit margin caused by market conditions, rather than an actual increase in physical output.

Features of Construction Productivity

We will conclude here by presenting key features of the organisation of the industry and the production process which need to be taken into account when analyzing productivity measures in construction. Construction is assumed by many to have 'low productivity' (*sic*) (Briscoe 1988: 282). The discussion of single factor productivity measures above has shown that it might, more accurately, be described as a labour intensive process. Bishop (1975) has presented a detailed analysis of labour productivity at construction project level which still has much relevance today. A key constraint to increases in labour productivity is that there are many discontinuities in the work involved on construction sites. On large sites gangs may follow each other sequentially from one location to another each carrying out a specific task. For example, fixing reinforcement, erecting formwork, pouring concrete, striking formwork. This creates problems of maintaining resonance between gangs, leaving sufficient headway for following and preceding gangs.

There are problems of tolerances, and, for example, high output leading to low quality and the necessity for following gangs to have to take remedial action before commencing their own tasks. This is the realm of construction management, which is concerned with the technology, the design and the organisation and motivation of the people on the site. Hard physical work, discontinuities between tasks and between projects and the question of ensuring that work is meaningful and that employees retain self esteem and reasonably comfortable working conditions are some reasons why at least one writer has questioned the whole basis of conventional productivity analysis as applied to construction projects (Lewis, 1987).

Portfolio Management

In addition to the capital invested in plant, offices and equipment, the working capital represents the investment of the firm's owners. The profit achieved on each project represents the return on this capital. The owners of the firm are interested in the overall rate of return achieved in a given time period over the range of projects or 'portfolio' of work carried out by the firm. The objective of the owner is to choose from the available set of 'investment opportunities' the set which best satisfies his or her objectives. Investors prefer high returns, in general these are unfortunately associated with high risk.

A decision criterion which maximises return, will therefore also maximise risk. The investor is usually willing to reduce risks to an acceptable level by choosing projects where both the risks and the returns are lower. In essence, the problem is one of establishing the optimum combination of risk and return among the set of projects which the investor chooses. It is on the path to this decision that portfolio theory is of relevance.

Portfolio theory was pioneered by Markowitz (1959). The objective of portfolio theory is not to identify the optimum portfolio but rather to identify the so-called 'efficient set' of investment alternatives. This being done, the investor will then choose from the efficient set, the set which most satisfies his or her objectives. Markowitz' central contribution to the then existing literature was to suggest that the investor was not merely concerned with maximising the return, for this would lead to a minimal diversification into the two or three alternatives offering the highest returns. On the other hand, he pointed out that investors diversified widely in order simultaneously to minimise their risk while maximising their return. Put more succinctly, they attempted to maximise their return for a given level of acceptable risk. Thus, it became clear that the correlation between investments was crucial. A pair of investments which move up and down together are said to be positively correlated. A pair which move in opposite directions are negatively correlated. Thus, diversification is only sensible in terms of risk reduction if the marginal project is negatively correlated with the bulk of the existing portfolio.

The investor's objective is to achieve a portfolio with the desired profile of risk and return. According to Vergara and Boyer (1977: 25), 'The return and risk characteristics of the portfolio are a function of the risk and return of each security, the correlation between securities, and the proportion of the total portfolio invested on each security', In essence, to reduce risk, it is necessary that the investor avoids a portfolio which is highly correlated. Selection of a portfolio follows three steps, firstly identify the characteristics of each possible investment, secondly identify the characteristics of each portfolio, thirdly select the required portfolio.[8]

Notions of portfolio theory added to the system of sealed bidding for construction projects have major consequences for the application of orthodox neoclassical theory to construction. Efficient management of portfolios, with diversified and negatively correlated risks, may be at least as important as productive efficiency. It is by no means certain that the least-cost, productively efficient firm will be the one to survive. Over time, the most profitable firm may be the one that is best at portfolio management and not the one which is most technically efficient (Ball 1988: 97). Given that return on capital is a crucial criterion for the continued existence of any firm, it should be no surprise that a further implication of portfolio theory is that the pressure for innovation will be at least as much in the non-technical, financial and accounting dimension as in the technical, production dimension of the firm's operations.

Examples of this are provided by the innovations in project financing which occurred in the 1980s during a period of international credit crisis for major projects. Where there is no finance, there is no project. When traditional sources of credit and finance dry up, or are unavailable to particular (high risk) borrowers, then it becomes necessary to search for alternatives.

The main sources of new credit in the 1980s were the projects themselves. If, for example, there is sufficient shipping business to generate a profit from the operation of a new port, then there should be a viable project. In commercial and bilateral projects the use of 'Build Operate Transfer' (BOT) contracts is a good example of new approaches to the sharing of credit risks. In this type of contract, the contractor, the operator and the financiers all have pecuniary interests in both the outcome of the construction phase and the overall project viability as affected by the postcommissioning revenue.

During this period there also grew up the use of multilateral financing schemes seeded by, so called 'B' loans from the World Bank. Under this scheme the World Bank becomes a co-financier with groups of commercial banks and bilateral export credit agencies. The advantages of this scheme are that risks are spread more widely, and that the presence of the World Bank was thought to decrease the likelihood of a default and therefore increase the chances of commercial bank involvement (Selsdon and Palmer, 1988). These innovations were designed to divide up and share economic, technical and political risks. They are financial and accounting innovations. They involve high transaction costs and are unrelated to the technical, productive efficiency of the contractors and construction consultants involved. Be that as it may, without these financial innovations many projects would not otherwise have been realised.

Strategic Management of Firms: An Example

By 'strategic' here we mean 'in the medium to long term'. Dictionary definitions of strategy refer to artifice and finesse in planning battles, as well as generalship in gaining decisive advantage. We shall be more concerned with the latter than the former.[9] We shall consider the strategic management of the firm by observing what happened to the UK construction industry in three distinct economic environments. These are, a stable environment, an environment characterised by unexpected shocks and an environment which necessitated restructuring of the industry.

The period 1960–1990 in the UK industry is an excellent one for studying how firms have operated during different types of economic environment. It is also especially interesting because the period from 1979 to the time of writing (1989) was one of continuous government by one party with a sustained and largely consistent economic and political ideology. We shall first briefly describe the period in terms of construction industry activity and the economic and social environment. Second, we will consider the implications of these changes for the management of the firms. Third, we will look at the results, what actually happened to the firms and to the people who worked in the industry. Finally, we shall assess the extent to which it is possible to learn any general lessons about the operation of construction firms from the study of this period.

The dates of 1960 and 1990 are slightly arbitrary. Any start date during the post war recovery period could have been chosen. 1960 is convenient because it is relatively easy to collect statistics from this date which run to the present. The end date is governed primarily by the time of writing.

Figure 8.6 illustrates the real change in construction output for the period. The first pattern which emerges is that the period up to 1969 was one of stability and steady growth. From 1969 onwards there is much more instability. We have chosen to divide this into two. The period from 1969–1980 demonstrates volatility and upheaval. There was the first oil crisis and then the UK property collapse resulting in two recessions in the construction industry, the second of which caused output to fall to the level it was at in 1964. The third period begins with the major recession of 1980–81 which caused the output of 1981 to fall to the level of 1963 in real terms. Thus, in period 1 there was steady growth and relative stability. In period 2 there was upheaval, sharp, unexpected and unprecedented change, large oil price rises affecting investors confidence, high inflation and a property crash. In period 3 there was a very sharp but quick recession, which set output back by eighteen years, followed by eight years of steady growth. The depth of the recession was such that output did not again reach its 1979 level until part way through 1986. The end of period 3 was marked by the onset of a new recession in 1990.

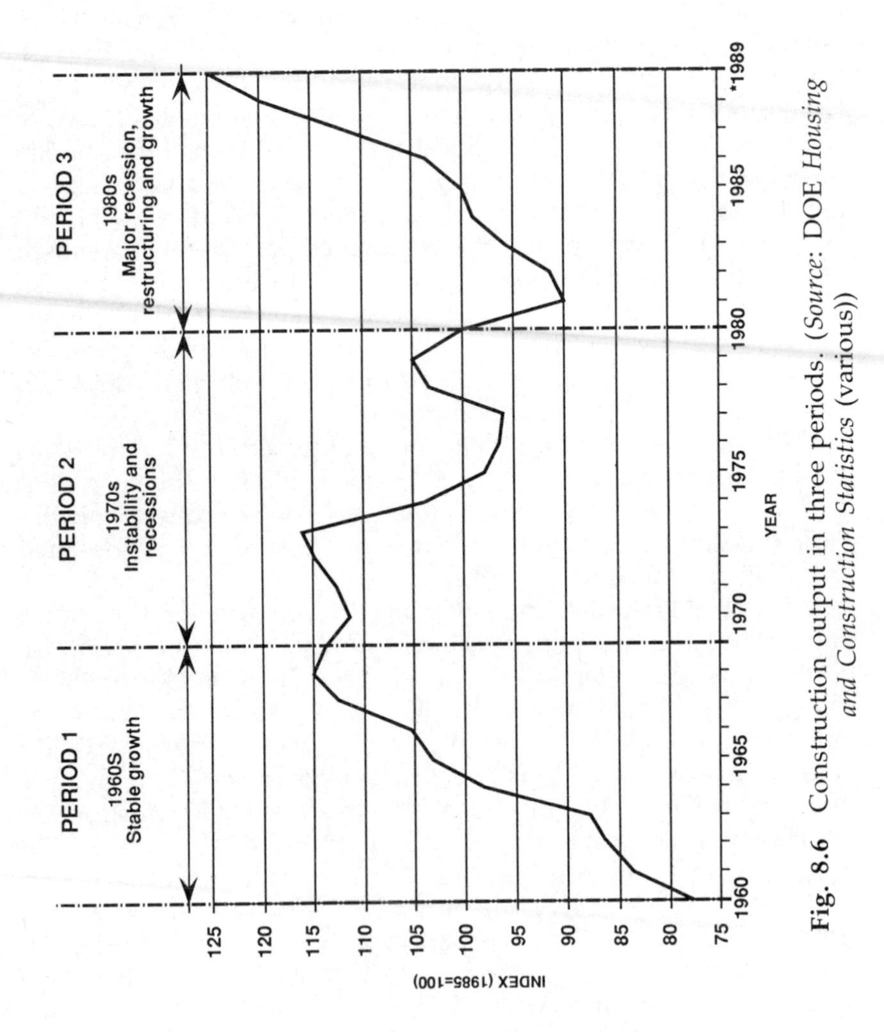

Fig. 8.6 Construction output in three periods. (*Source:* DOE *Housing and Construction Statistics* (various))

The economic and social environment of the three periods is also quite distinctive. The 1960s were the 'never had it so good' years, a time of growth and development, of economic and social optimism. The 1970s saw a series of economic shocks and the growth of unprecedented inflation and unemployment. By the time the 1981 recession had been fully felt a new government had been in power for two years with a political and economic ideology which was to set the tone for the remainder of the decade. This involved the privatisation of large parts of the public sector, the introduction of 'market economics' to areas of public services, tax reductions and tightening up of the rules for eligibility for social welfare benefits.

Lansley (1987), in a study of corporate strategy in the construction industry, uses Eppink's (1975) characterisations of environmental change to assert that different management styles were needed to respond to the different environments of the three periods. The stable period enabled firms to survive and grow by formalizing and learning from their past experience. The stable economic environment meant that management could afford to be introspective, focusing on means of increasing internal efficiency. There were no permanent changes in the relationship between the firm and its environment. The nature of environmental change was *operational* and could be coped with by preprogrammed responses and attention to technical efficiency. In the 1970s the problems encountered could not readily be handled on the basis of experience but required creativity and ingenuity. In order to respond to the environment, firms required 'flexibility'.

According to Lansley (1987: 147) a corporate, 'people oriented' style was needed. A style which recognised that workers in the lower levels of the organisation had detailed knowledge of the markets and environment and thus, had an important role to play in identifying strategy for the firm in a time of *strategic* (according to Eppink, 1975, unexpected or unprecedented) change. The 1980s were, according to this approach, characterised by *competitive* change. That is, change which is long term and which involves restructuring of firms, and the industry, and changes in the way the industry relates to its clients. Thus, in this period the firms have had to grow accustomed to lower levels of demand, to higher expectations of more specific types of service from clients. This requires managers to readdress issues of task efficiency while giving increased attention to control within the organisation and to the design of specific services for specific client needs.

All this is very conceptual of course, what *actually* happened to the industry, to the firms, to the people who work for them, during these three periods? Let us consider the organisation of the industry and its relations with clients, the structure of the industry's firms and their levels of profit, and finally, the conditions of the employees.

The nature of period 1 was such that very little change took place in the industry. The 1960s were a time of growth, firms and fortunes grew, workers enjoyed the benefits of 'full' employment. Both consultants and contractors became relatively powerful. Because there was no shortage of projects it was difficult for clients to exert any pressure on the industry to revise its procedures. The 'traditional' method of procurement was the most common. Under this system the client had separate contractual relationships with the architect, the structural consultant, the services consultant, the quantity surveyor, the main contractor and, on occasion, with specialist and trade contractors and suppliers. With no shortage of clients or projects it was natural that sentient domination should cause each of these participants to look after their own interests. It became very difficult for the client to ascertain liability for defective or late work. Only the most sophisticated or experienced clients could, on receipt of legal advice from the construction legal industry, deduce who was responsible for what. A seminal report by Higgin and Jessop (1965) identified two serious problems in the industry. Namely, confused notions of roles and responsibility, and difficulties in communication among the members of the building 'team'.

It was not, in fact, until the early 1970s that the dissatisfaction of clients with the services of the industry began to have any real result. The result of this dissatisfaction was the eventual emergence of new forms of construction procurement. New developments were resisted by the industry as may be seen from the titles of some articles in the technical press at the time. Consider for example, *If We Must Have Package Deals* (Jefferies-Matthews 1970); *Package Deal Threat* (Gwynn 1968); *The Consultants Case Against Package Deals* (Lowe 1970). The net result of this was that the industry began to take serious steps toward the development of new procurement methods such as package deals, design and build, project management, and management contracting (Franks 1984).

Key features of all of these developments were, a reduction in the number of interfaces between the client and the industry, a clarification of responsibilities of members of the building team, increased attention to the *control* of time and cost, an explicit separation of the role of 'management' and a consequent reduction in the power and role of the architect. The industry's reactions were, however, not enough for the clients. The British Property Federation, an association of clients who were frequent users of the industry, brought out their own model form of building procurement (BPF 1983). That a group of clients should feel so dissatisfied, they felt it necessary to commission their own research and design a system of procurement which suited their needs better than any of the options being made available by the industry, was, and still is, a deeply embarrassing event for all members of the construction industry.

By the end of period 3 the 'new' forms of procurement were no longer new. The industry's clients had firmly wrested power back from the industry, via lower demand, and had successfully forced the industry to increase its performance on delivery times by introducing methods of 'fast tracking' projects.

Let us consider now the firms themselves, in terms of their profits performance. As the product of construction is, usually, sold before it is produced, we may take the output of any given year as an indicator of the demand for, say, the previous year. The falls in output in the years 1970, 1974–77 and 1980–81, indicate severe drops in demand. Maintaining fixed costs while producing lower output would lead one to expect that profits should fall. Collecting empirical data for the study of profit levels in the industry is difficult. Ball (1988: 126–52) using data from the Bank of England Quarterly Bulletin of June 1982 and from Williams (1981), shows that, with the exception of one year (1973), for 1961–1981 rates of return in construction were consistently higher than in manufacturing. There was though, a general drift downward in profit levels during the three periods. The point is though, that profits were not affected as disastrously as output, or as one might have expected.

This feat was achieved in two ways. By diversifying into overseas work and property development, and by squeezing down input costs. We shall consider only the latter, as it had a major impact on the people who worked in the industry. The most important feature of the massive restructuring of the industry which took place in the 1980s was the way in which large firms divested themselves of permanent employees and

Table 8.4 Number of contractors in the UK registered as having 1 employee, 1978–88

Year	*No. of contractors*
1978	28 551
1979	30 197
1980	36 549
1981	40 580
1982	55 498
1983	64 585
1984	71 386
1985	72 896
1986	79 946
1987	79 354
1988	83 484

Source: *DOE Housing and Construction Statistics* 1978–88, London, HMSO

sharply increased their use of sub-contractors, specialist trade contractors and the self-employed. Empirical evidence for this is presented in Table 8.4.

Contracting firms are able to increase their flexibility by using forms of labour which may be quickly hired and fired. While they are working, the self-employed may earn more than they would have done had they been directly employed. However, they need to allow for periodic episodes of unemployment and they become fully responsible for their own pension and health provision. They have to file their own tax returns. Thus, the firm's costs are further reduced by the costs of administering directly employed workers. The greatly weakened position of the construction workforce has been analyzed in detail by Ball (1988: 114–25, 198–204).

Let us, finally, attempt to draw some general conclusions from these observations of construction firms in three types of economic environment. The period of stable growth and few surprises for the management resulted in rather introverted firms who continued to supply the service as they always had done, despite growing evidence of time and cost overruns, and client dissatisfaction.

During the instability of period 2 the firms started to reduce the number of directly employed workers. During period 2 and 3 there was a steady fall in construction employment and a steady increase in the number of self-employed. Period 3 required firms to accustom themselves to a lower level of demand. This they managed to do by restructuring both the firms, the services they offered and, it must be said, their attitude to clients. In all, the result of two decades of recession and recovery is that the firm's profit levels fell but still ran ahead of industry in general, while the employees found a reduced number of available jobs and frequently reduced stability and conditions in the jobs that do materialise. Firms in a labour intensive industry had successfully reduced their input costs and had been forced to develop new ways of meeting the needs of their clients.

Notes

[1] The problem of actually measuring gross operating profit and financing costs of relatively lengthy construction projects with phased payments in a fluctuating economic environment has not been treated well in the literature of building economics. A rational framework, taking into account the time value of money is presented in Au and Hendrickson (1986).

[2] There are many critics of the theory, see for example, Fitzroy and Mueller (1984), Malcolmson (1984). A lucid description of Williamson's work is given by McGuiness (1987).

[3] 'Overseas' projects or, more precisely, projects anywhere in the world, for which there is international competition, constitute a special case of price determination. There appears to be a large degree of non-price competition. Governments frequently spend public money in insurances (export credit guarantee schemes), financial arrangements, overseas 'aid' with binding agreements on where the country being assisted must purchase certain goods and services, and pre-bid intergovernmental contact in order to ensure that their own national contractors are in a favourable position to win export orders. For more detail on this specific field see Strassman and Wells (1988), Institution of Civil Engineers (1988) and Raftery (1985a).

[4] There is, albeit a bit dated, some empirical support, in the UK at least, for this decision. In a survey carried out for the UK National Economic Development Office it was shown that 'selective tendering' was the single most frequent method of appointing the main contractor used by public

clients, NEDO (1975). Public clients need to ensure public accountability for their expenditure and so they have a greater propensity to use 'open' tendering. Having shown that even public sector clients use selective tendering most often, it seems safe to assume that private sector clients will use it at least as often and probably more so.

[5] There is a conventional wisdom in the industry that to increase the number of bidders would lead to increased prices in the long run, due to the need to recover the cost of preparing unsuccessful tenders from the successful projects. Skitmore (1988) has pointed out that this wisdom is as yet empirically untested.

[6] *Pareto* optimality and '*Pareto's* distribution' are much misunderstood in construction. The so called '*Pareto's* distribution' in no way implies that 80% of the cost of a project is accounted for by 20% of the items in a bill of quantities, as is thought by some in this field!

[7] The figures were presented in the context of an analysis of supply-side change in the UK economy in the 1980s presented in the *Economist*, 3 December, 1988: 103–104.

[8] The applications of portfolio theory in construction are presented in detail by Vergara (1977) and summarised in Vergara and Boyer (1977).

[9] More detailed studies of business objectives, strategy and competitive advantage may be found in Ramsay (1989) and Male *et al.* (1989).

Chapter 9

Price Determination for Construction Projects in a Mixed Economy

Introduction

The aim of this short chapter is to present an ordered examination of the main factors which influence the price of construction projects in a mixed economy, in this case, the United Kingdom. Thus, in passing, we will evaluate the extent to which the orthodox mainstream model of price determination holds true in the context of the construction industry. First, we will review the orthodox model of price determination. Second, we will briefly review the supply conditions of the industry. Third, we will discuss, what some economists might term, 'frictions' in the price determination process. These will include the actual organisation of the industry, its institutions and its markets, the methods of appointing contractors, contractors pricing behaviour, the concept of risk/reward ratio. Finally, we will conclude by proposing an alternative model of price determination in the construction sector.

The Orthodox Model of Price Determination

Based on certain assumptions about human behaviour, the nature of a 'perfect market', and, of course, assuming that all other things are equal, mainstream economic orthodoxy holds, that in a perfect market the price of a good will tend towards an equilibrium point where demand is exactly balanced by supply. In other words, if more of the good is supplied to the market, then the price will drop as suppliers compete to clear their stock. At lower prices some firms will go out of business and there will be a tendency to return to the initial equilibrium point. If, other things being equal, demand for the good increases for some reason then buyers seeking more of the limited stock will bid the price up. The possibility of high profits will encourage more firms to produce. Increased competition among firms will eventually force the price back to its equilibrium point. The price of the good will settle at a point where the market 'clears'. Further, the market has cleared due to movement in prices, and not to any other factor such as fluctuation in production. The amount that buyers wish to buy is equal to the amount that sellers wish to sell.[1]

The essential features of this model are thus; a set of assumptions and constraints, the notion of demand and supply schedules, and finally, a conceptualisation of the firm's output and pricing decision. These are all well known but it is worth rehearsing them briefly here.[2]

The first assumption, of what has been termed the 'equilibrium price auction' model, is that the economic man on the Clapham omnibus is a rational individual whose overriding objective is to maximise his welfare, via a concept called 'utility'. Utility, as we have seen, is the satisfaction or pleasure gained from the consumption of a good or a selection of goods. Marginal utility diminishes as consumption increases. A champagne loving consumer gains less utility from the second bottle than the first. If the price of a good falls, the consumer may purchase more, but at a decreasing rate. If the price of champagne falls to the level where the consumer could easily afford to buy three bottles, the consumer may chose to maximise utility by not buying a third bottle but by, instead, buying a patent hangover treatment.

Utility, therefore, is a rather interesting catch-all kind of concept which may be used to rationalise seemingly irrational behaviour. This is very useful when the fundamental assumptions of the economic paradigm are based on rational behaviour. When humans behave in an irrational way, should we put this behaviour through, what amounts to, a transformation, via the concept of utility, in order to render it 'rational' for the purpose of our model? Or should we revise the model to produce one which more realistically captures the essence of the rich, and sometimes erratic, variety of human behaviour?

The second assumption is that the transactions take place in a perfect market. The perfect market is a theoretical construct whose main features are briefly as follows. There are large numbers of buyers and sellers so that each firm produces a small proportion of total output and no one firm can exercise strong influence on the selling price. All buyers and sellers have perfect information concerning the prices and quantities sold in all transactions. There is zero friction from transaction costs and entry and exit from the market can occur rapidly and incur no costs.

The third important assumption is that both firms and consumers can make smooth marginal adjustments in their economic decisions in order continually to maximise utility and profit respectively. We will return to this point below.

Let us turn now to the notions of demand and supply. Demand, as we have seen, refers to the amount of a good which buyers are able and willing to purchase at each price in some conceivable range. On the assumption that, other things being equal, at higher prices consumers will buy less and at lower prices they will buy more, we arrive at the familiar demand curve which slopes downwards to the right. Similarly, the supply schedule illustrates the relationship between the market price and the amount of a good which firms are willing to supply.

How does the firm decide on its output and on the price at which it should offer its goods for sale? The supply curve slopes upwards to the right. As price falls, firms will be willing to supply less of the good. Conversely, firms will be willing to go on increasing production as long as it is worth their while. In other words, as long as the marginal revenue generated by each additional sale exceeds the marginal cost of producing that last unit for sale. In theory, the firm will keep increasing production while its marginal cost is rising (assuming that all output can be sold). It is in the interest of the firm to hold production at the point where the marginal revenue generated by the last sale is equal to the marginal cost of producing it. Clearly, if the firm went on increasing production, the marginal cost of each extra unit would be more than the marginal revenue generated from the sale and total profit would fall. Thus, the point where marginal cost equals marginal revenue is the point of optimum profit. To produce any more, would be to incur a loss, as the cost of producing the additional unit would exceed the price which could be had for it. To produce any less, would be to fail to wring the last drop of profit out of the market. This is illustrated in Fig. 9.1.

Clearly, the basic model needs to be adjusted to take account of the fact that the majority of real world markets have many 'imperfections' and that other things are rarely equal. Thus, there are variants of this model for monopoly (one firm dominating) and oligopoly (small number of firms dominating). It is important to note though, that the fundamental starting

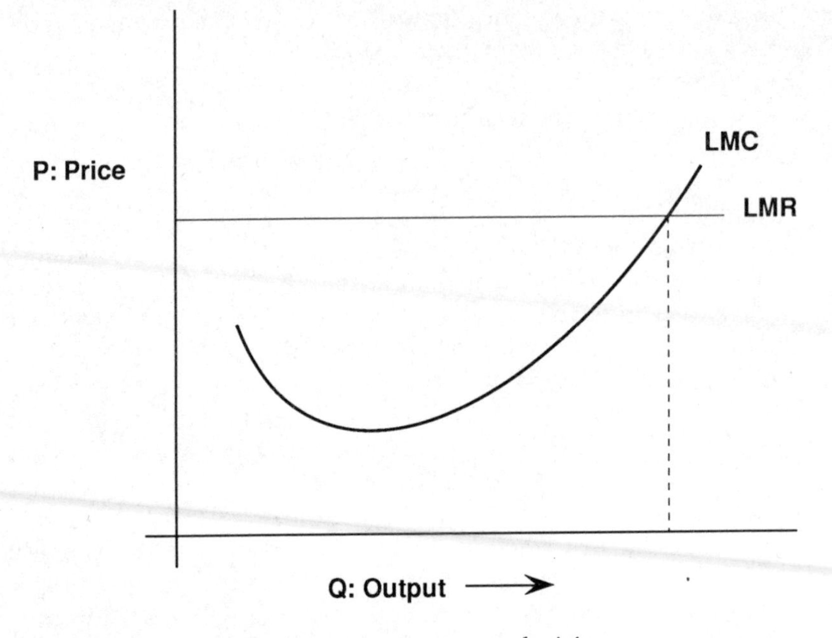

Fig. 9.1 Firm's output decision

point is the equilibrium price-auction model with assumptions of rational optimising behaviour.

Supply Conditions

Behaviour of the Firm

Construction firms, in common with other types of firm operating in the economy, may hold a number of, sometimes conflicting, objectives. In order to satisfy the owners or shareholders, they need to generate a profit which represents an acceptable rate of return on the money invested. The personal objectives of the owners or of the managers may lead the firm to have growth of turnover as an important objective. If demand in the firm's chosen type of work is falling, then survival may become the important objective. In either of the latter cases the firm may then be forced to reduce the priority attached to generating a profit in the short term, in order to develop into new markets. Many firms will attempt to achieve the same objective by product differentiation and by other forms of non-price competition.

The 'Market' for Construction

As we have seen in chapter 5, the market for construction work may be usefully segmented by geographical region, type of firm and work specialisation. While there is 'ease of entry to the market' due to the low working capital requirements, certain other perfect market criteria are not met. For example, construction projects are often awarded to contractors in the context of small numbers bidding and bargaining. Sealed bidding procedures ensure that there is minimal knowledge of other sellers' offer prices.

Market Imperfections or Observed Constraints?: Real Features of the Pricing Decision

Impossibility of Smooth Marginal Adjustments

The most commonly used method of price determination is some form of bidding. Approaches to this are described elsewhere.[3] In terms of the theory, the contractor (the supplier) bids for the privilege of being awarded the contract to supply the building for a specific price. There is usually a small number of bidders, i.e. less than ten. Similarly, the perfect information criterion is not met, as the bids are sealed. The bidders do not know the prices being asked by their competitors. If the bidders have been selected by the clients advisors it is probably safe to assume that, for all intents and purposes, the product is homogenous, i.e. the competition is truly on price alone. Or is it? Interestingly, most sensible client advisors will clearly state that they will not be bound to accept the lowest or any tender. Why might this be? Might it be the case that price competition is helpful only within certain parameters? Might it be the case that these clients are explicitly recognising that the products may not be homogenous?

Construction projects are often very large when compared to the size of the firm, due, as we have seen, to the system of interim payments and the multiplier between working capital and project size. Consider a firm working at less than its capacity but, more or less, covering its fixed costs. The winning of a new project may not represent a smooth marginal adjustment upward in production. For all but the very largest construction firms in most countries in the world, the likelihood is that it is a relatively large stepwise jump in the number of ongoing projects. It is impossible to imagine even the most sophisticated construction firm having the least idea of its marginal cost and revenue curves. The smooth curves of Fig. 9.1 now need to be transformed into stepwise functions. This is rather more than a 'market imperfection'. The problem simply emerges from the inapplicability in construction, of a model based on industrial mass production.

In practice a method of mark-up pricing is used. As we have seen, this has led to a full literature on bidding models based on the assumptions that in perfect competition, in equilibrium, all firms should have the same costs. Hence, firms with high costs would be inefficient and, over time, would go out of business. Clearly, the higher the mark-up the greater the potential for profit but the lower the chance of winning the project. A contractor who desperately wants to win a project may be willing to use a lower mark-up and *vice versa*. In practice, the cost estimate for a project may be produced by a team of estimators, this estimate will then be given to the managing director or some other senior person, or group, to decide on an appropriate mark-up. If the firm actively wants to win the project the essence of the task is to arrive at a mark-up which is sufficiently low to beat the competition and win the project, while at the same time, sufficiently high to make a reasonable profit. In this chapter we are going to focus on the factors affecting the level of the mark-up. In order to do so we must first dispose of the cost element of this model.

Costs arise as a result of the use of resources. In the case of the construction project, the work is to be undertaken at some future date and thus the costs are *estimated* costs. The contractor's costs will include the following; site labour, materials, equipment (owned or hired), insurances, headquarters costs and bank charges. These costs will be influenced by the location and technical content of the project, its size, shape, level of services provision and so on. These technical issues are quite well understood individually but they are also highly inter-dependent and it is difficult to derive general relationships. The increasing use of Computer Aided Design (CAD) systems incorporating cost models will mean that project appraisals which take account of these inter-dependencies will be easier and quicker to produce. It goes without saying, that although the contractor may only be interested in estimating the capital cost of the project, the building owner or user will of course be interested in the life-cycle cost.

The technical and cost modelling issues are discussed by Ferry and Brandon (1991), Stone (1983) and Raftery (1985).

The remainder of this chapter will be concerned with the factors which affect the level of the mark-up.

Risk/Reward Ratio

The prices of construction projects are usually agreed in principle before construction begins, sometimes even before design is fully complete. All forecasts are subject to uncertainty and construction programmes and cost estimates are no exception. Furthermore, construction is an inherently risky process, each project is unique, ground condition surveys are often inadequate, design details often need to be revised when construction begins. In most developed countries nearly 50% of construction work takes place in existing buildings of variable age and state of repair, with a consequent increased uncertainty as to the precise nature and extent of the work necessary to rehabilitate or repair to a specific standard.

Investors are, to some extent, risk-averse. Where risks are higher than normal they will expect some additional profit to offset their risk aversion. The greater the potential for things to go wrong, the greater will be the reward required for carrying such risk. Some risks are insurable, and in this case the additional cost could be taken as the cost of the insurance premium. In cases where the risk is not insurable the firm will need to make some assessment of its risk/reward ratio in order to calculate its own 'intra-firm' risk premium. The concept has particular applications in the types of procurement method and legal contract used on construction projects.

Method of Procurement/Contract

Finally, we focus on the point where all of the factors come together to determine the price for projects. The link between the supply side and the demand side of the industry is made through the method of project procurement used by the industry's client. The method of procurement affects price determination mainly through the way it distributes risks between contractor and client. The determined price is also heavily influenced by the state of the market and the contractor's desire for the project.

Franks (1984) describes seven types of procurement, all but

one of which are in common use in the UK industry. They may be arranged into two groups, a traditional tender based appointment of contractor and a group based around some sort of management for a fee. They are as follows:

- The traditional system.
- Package deal and turnkey contracts.
- Design and build.
- Separate contracts.
- The BPF system.
- Project management.
- Construction management.

The detail of these are not relevant here but suffice it to say, that they involve varying distributions of responsibility for design and construction and varying allocations of risk between the client and the various industry participants in the project. The first group are based on a tendering approach, usually selective tendering. The BPF (British Property Federation) system and the 'design and build' place high levels of responsibility with the contractor, who will, logically, require a premium for this, over and above the additional cost of the design work. This does not mean that they necessarily give bad 'value for money'. It does mean though that the risk/reward ratio will be carefully assessed by the firm.

In order to save legal time and expense, standardised forms of contract are used. However, these have become more and more specialised and the number of 'standard' forms is increasing. Some forms of contract contain price escalation or fluctuation clauses. These allow the contractor to claim for any price increases (or reductions) occurring after the contract was signed. This has the effect that the client pays the *actual* escalation on the project. Other contracts require the contractor to give a, so called, firm price which includes an allowance for inflation. In this case the contractor carries the risk. In periods of volatile inflation a risk averse forecast of inflation will be used. Hence, on the outturn of the project the client may have paid more for inflation than if he or she had agreed to adjust for it accurately after the event. Many clients however, regard it as worthwhile to have a firm price at the beginning even if it actually causes a higher price. They regard it as a worthwhile benefit to have certain knowledge that the construction price will not escalate.

Finally, we should include a short discussion on some of the problems of tendering. It has been said that the lowest tender is often the one with the biggest error. In selective tendering contractors are *invited* to bid. Only those who have shown that they are capable of undertaking the project and who have adequate management skills and financial and material resources will be selected to bid. Even in this case however, distortions arise. Contractors who happen to be on a tender list but who do not particularly want a specific project may rarely disclose this to the client. Instead they submit a so-called 'cover price'. This is a price which is far too high for the project and which will probably not be the lowest tender but which, in the unlikely event they are awarded the project, will be so high as to make it worth the firm's time to undertake the work even if their resources are fully committed elsewhere.

Market Conditions

Thus far we have rather artificially disentangled many factors affecting price determination. They are interdependent and complex and their resultant effect may be observed in the prevailing market conditions. The net result of the factors we have discussed so far, is that the notional equilibrium point is never reached. This may be seen by considering the differential movement of input costs and output prices. Figure 8.6 plots the overall output of the construction industry in England and Wales 1960–1989. Consider the period since 1980. The severe downturn of 1980/81 was followed by six years of strong growth. Figure 9.2 plots the movement of input costs and prices charged by contractors over the same period. Clearly, as demand for construction increased, the price charged by contractors increased at a much faster rate than the rate of increase of their input costs. Interpreted loosely then, the model does help to describe the behaviour of the market.

Theoretical Inadequacies of the Orthodox Model

Thus, loosely interpreted, the rational, optimising, equilibrium price auction model provides a useful analysis of market behaviour. However, real world observations suggest that the minutiae of the model are not as close to reality as we might wish. Furthermore, some of the fundamental theoretical bases of the micro economic model have been thrown open to doubt by the work of many economists in the past few decades. This work

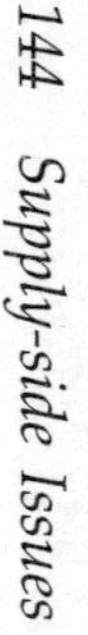

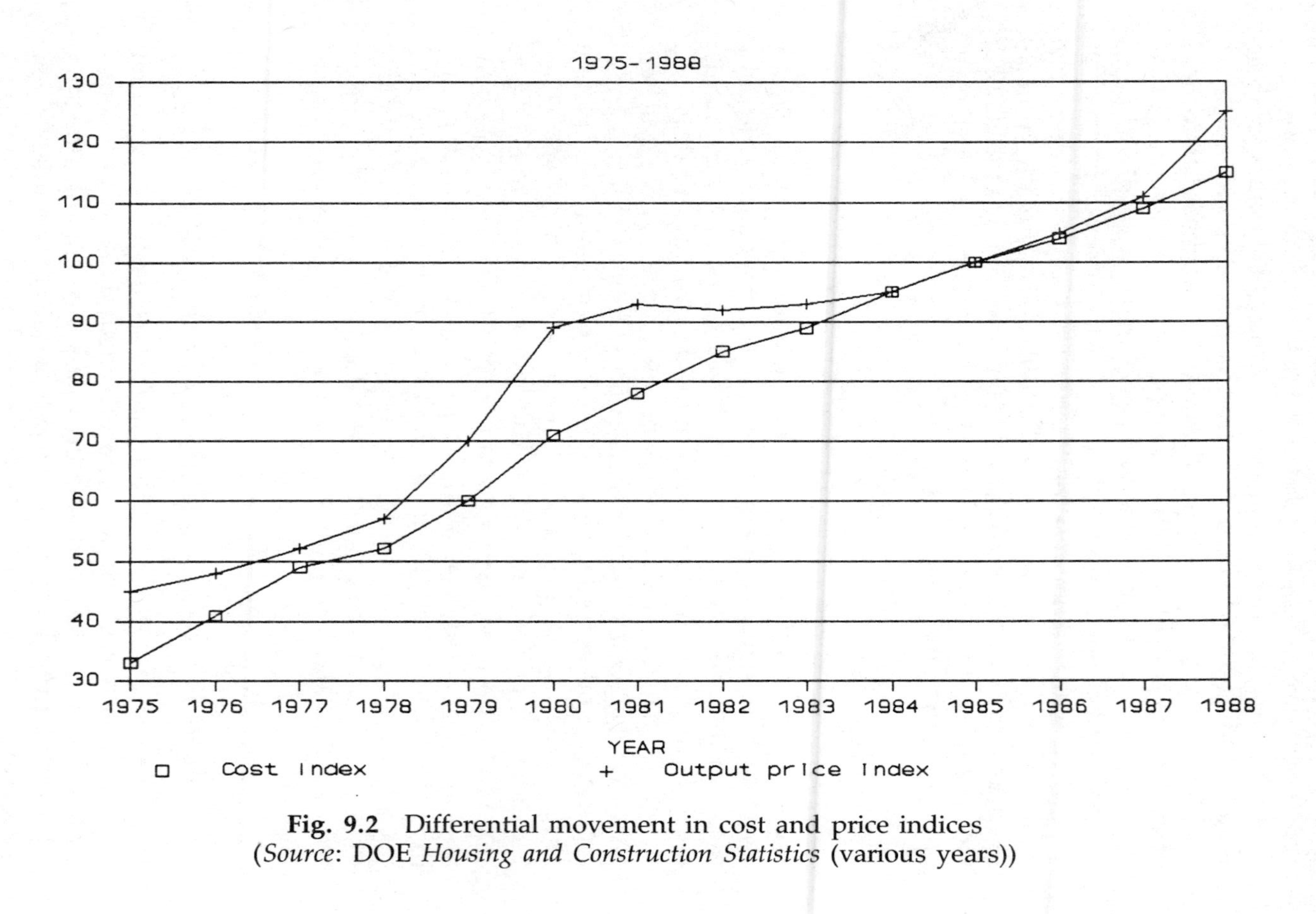

Fig. 9.2 Differential movement in cost and price indices
(*Source*: DOE *Housing and Construction Statistics* (various years))

has been assembled in a seminal book by Andrew Kamarck (1983) from which the following paragraphs draw heavily.

Turning first, for example, to the notion of movement along demand and supply curves. It is not safe to assume that just because we can establish a relationship between price and quantity demanded over a small range of prices and quantities, it is then possible to deduce the relationship over the whole range, (Georgescu-Rogen, 1987). Further, in the words of Kamarck (1983: 94), 'Economists have been trained to assume that people always optimize'. This is the basis of microeconomic theory. It is therefore, further assumed that firms will optimise by maximising profit in the manner outlined earlier. This further assumption is based on the rationalisation that if one person optimises then groups of people together will similarly optimise. There is no evidence from behavioral science to support this simplistic assumption. In fact, such evidence as there is, tends to suggest that group behaviour is quite different to individual behaviour. This notion, that inter and intra-group behaviour is different, is the basis of Liebenstein's (1976, 1977, 1978) theory of 'X-inefficiency'.

X-inefficiency is the shortfall between the theoretical optimum and what actually results when various groups of workers in the firm negotiate with each other, in the formal and informal structure of professional and personal relationships, and act generally, but not universally, toward the achievement of the firm's objectives. Thus, X-inefficiency takes for granted that the theoretical optimum will never be achieved.

The notion of X-inefficiency at group (firm) level is supported at the level of the individual by Simon's (1979) behavioural theory of rational choice. This states that people do try to make rational choices but that they rarely have perfect information and usually settle for the 'good enough' rather than the 'best'. So, for example, if we are about to buy a used car of a specific age and condition, we search through the local dealers and the classified advertisements for the best price. We are fully aware that if the search were expanded to include every local newspaper in the country it would be highly likely that we would find the optimum (lowest) price. However, we decide that in order to make the time and effort manageable we will restrict ourselves to, say, one or two adjacent areas. Under conditions of bounded rationality we find a local optimum which we regard as good enough. In other words, neither people or firms behave optimally, and even if individual people did, it is unlikely that firms would.

Finally, Scitovsky (1976) has shown that the very concept of *utility* is a one dimensional oversimplification of human behaviour. He proposes that rather than assuming that people have a simple desire to maximise satisfaction or utility, we should consider that there are at least three motivating forces behind human behaviour. These are, the drive to relieve discomfort, the desire for stimulation to relieve boredom and the desire for the pleasure that can accompany and reinforce both of these. This approach has clear foundations in behavioural science and psychologists (but, unfortunately, not economists) have known this for a hundred years. The problem with X-inefficiency, bounded rationality, three dimensional utility and even simple ideas about group behaviour theory, is that they add up to the unpleasant (for economists) reality that the world is a lot less rational and precise than certain economic model and theory builders would like it to be. A more realistic model of price determination would need to be at home with loose or fuzzy concepts, not feel guilty about lack of mathematical precision (physics envy) and be able to cope with erratic non-optimising decisions.

An Alternative View of Price Determination for Construction Projects in a Mixed Economy

Thus, we can deduce that when price changes do take place, other things are never equal. It follows, that these price changes result, not from movement along supply and demand curves, but from a continuous diagonal moving up and down to new curves. Add to this the reality of the non-optimising behaviour of firms and it is safe to conclude that the orthodox model has limitations predictive ability. Though, to be fair, it does indicate clearly that if the demand for construction increases or the supply decreases, then the price will increase. Similarly, if the supply increases or the demand falls then the price will decrease. The essence of the model is though, an equilibrium point which is never reached.

Figure 9.3 illustrates a different conceptualisation of price determination which is based on the concept of risk/reward ratio. The primary determinant of construction project price is the risk attitude adopted by the winning firm in the form of their risk/reward ratio as expressed through their mark-up decision. This risk/reward ratio is influenced by the general economic environment, by the number of close competitors, by

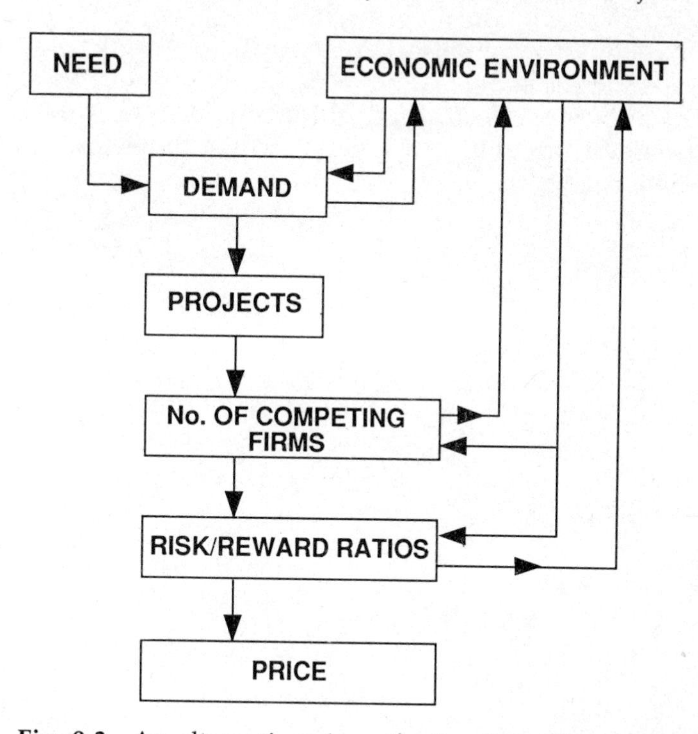

Fig. 9.3 An alternative view of price determination

the objectives of the firm and by the amount and quality of information available to the decision taker. Leaving to one side the general economic environment and the objectives of the firm, the number of competing firms is also influenced by the number of projects available. This has led one writer to suggest that, contrary to general opinion, the firms constitute the demand, and the projects the supply.[4] The number of projects available is influenced by the demand for construction which is a function of need on the one hand and the general economic environment on the other.

Notes

[1] It is important to distinguish between movement along a demand curve and movement onto a new curve. The latter occurs when other things are not equal, for example, if demand had increased because of a rise in disposable income. This would lead to a new demand schedule.

[2] The most lucid explanation of the detail of the model is

undoubtedly Samuelson and Nordhaus (1985).

[3] See Chapter 8.

[4] For a more detailed treatment of this point and of bidding generally (see Flanagan and Norman 1989 and Skitmore 1989).

Chapter 10

The Design Process

Introduction

It is usual in the construction industry for budgets to have been set and benefit-cost studies completed, long before the building has been designed. One consequence of this is that much energy is subsequently expended in ensuring that the developing design adheres to the budget. This is sometimes referred to as design cost control. Thus, an understanding of the design process is helpful for anyone who practices in this particular field of applied building economics. Specifically, we need to be able to draw some conclusions regarding those features of the design process, as it is currently understood, which may have an impact on the derivation and use of cost and price forecasting models. We need to be aware of the nature of the problem which faces the designer and of how he or she sets about the task of designing.

How is it possible to recognise a building which has a good design? Why do we regard some buildings as being well-designed and others not? A first attempt to understand the characteristics of the problem would be to listen to working designers' descriptions of how they have designed certain of their buildings. A work by Suckle (1980) did just that. Ten internationally acclaimed architects, each with major buildings to his credit, discussed their own processes of design and construction. This produced ten different descriptions of the design process which primarily outlined the most important of the personal concerns felt by each designer but added little to our knowledge of the general case. On the other hand, it is possible to read articulate descriptions of design methods and descriptive models of the design process by people who do not have acclaimed buildings to their credit. The suggestion would seem to be that good designs may be produced when the books have been thrown away but, as in so many other areas, you cannot throw the books away until first you have read them. This maxim provides some justification for the solution-orientated teaching of design which takes place in schools of architecture.

The process of design is notoriously difficult to study. (Whether indeed the term 'process' is applicable is itself a matter of some debate and is discussed below.) The difficulty is increased by the fact that the only evidence it is possible physically to observe is the end product, the building. It is important also to have some knowledge of the mental activities which take place during the period before the design problem is solved. To this end, the strategy of this chapter will be as follows: We will begin by considering the activity of design in terms of definitions, aims and conceptual views of working designers and other writers in the field of design studies. Second, we will examine some methods and models of design as they may shed some light on the process itself. Third, we will review the results of some empirical research on the design process. Finally, we will attempt to draw together some conclusions about the process of building design.

Design

Leaving aside Vitruvius, until relatively recently, the act of designing buildings or artifacts had not been very well documented. Much is known about architects and architecture in Greece and Rome, in the middle ages and in Renaissance Italy. (Kostoff 1977, Wittkower 1962) This knowledge concerns the views of the designers on ratio and harmony in the finished product, the building. Literature on the design and process itself began to appear in abundance in the early 1960s. The concern was with 'Design' in the global sense, encompassing inter alia, painting, crafts and all branches of engineering (Jones and Thornley, 1963).

While it is impossible (or at least extremely difficult) to ascertain the mental processes which take place when a designer is at work in the traditional sense, i.e. without the aid of design methods, there are however three points about which many writers agree. These have been outlined by Jones (1970) and may be summarised as follows:

(1) There are very often long periods when the person who is about to make an original work seems to do nothing except take in information and labour fruitlessly at seemingly trivial aspects of the problem. This is known as 'incubation'.

(2) The solution to a particular problem or the occurrence of an original idea often happens when at some particular point in time everything seems to fall into place. This is known as the 'leap of insight' or 'change of set'. Basically the problem is perceived in a new light and very often an apparently complex problem turns into a very much simpler one.
(3) The main enemies of originality are mental rigidity and wishful thinking.

This view of design owes much to the Gestalt school of psychology, prominent in the first half of this century (Koffka 1935). This school aimed to articulate some of the processes and representational structures which underlie behaviour and consciousness. However, current thinking and experimental results over the past few decades have greatly reduced the importance of this approach (Silveira 1971, Posner 1973, Chapman and Jones 1981). As the phenomena of incubation and change of set must be familiar to most people who have ever had to deal with a seemingly difficult problem (building design certainly qualifies) they are worth pursuing a little.

Incubation as a concept came into currency in the early twentieth century with the mathematician and philosopher Poincare, whose work was anecdotal and essentially derived from the introspective accounts of eminent scientists and artists. (Poincare 1929). He saw the conditions for creativity as:

(1) A period of conscious work, data assembled, problem defined etc., and some trials made at solutions.
(2) The unconscious works at useful and fertile combinations during this time and useless areas are inhibited.
(3) A hypothesis is derived which gives a fruitful direction. In his own words 'a period of preliminary conscious work always precedes all fruitful unconscious work'.

The implication of Poincare's work and of the first two points proposed by Jones was that following a period of intense work, the solution to a problem was likely to pop into the mind spontaneously after the problem had been set aside for a while.

This of course assumes that design is a 'problem-solving' activity. For the sake of argument we will assume further that in building design the particular design which is eventually produced is the 'solution' to the 'problem' as perceived by that particular designer. We should not assume that it is the only solution or that its accuracy is in any way objectively measurable, but merely that it is a solution to the problem as perceived by the designer. The further implication of both Jones and Poincare was that the solution was produced by an unconscious thought process. This mystical approach is remarkably similar to the concept of 'satori' or enlightenment in Zen Buddhist thought (Humphreys 1976).

Empirical evidence which apparently supports this was presented by Silveira (1971), who found that an interruption after a period of effort lead to an increase in the probability of solving a problem for both long and short interruptions, with respect to a control group. A further finding was that subjects did not return from the interruption with complete solutions in hand. It is important to know why this is the case, but this is not so clear.

Some possible explanations are:

(1) Incubation is a matter of rest. Rest gives a break and reduces fatigue.
(2) During incubation the subjects forget inappropriate sets and directions formed during the original strategy searching.
(3) The rest provides an occasion for additional practice on the problem.
(4) Incubation offers scope for a chance occurrence of an external event which completes the problem.
(5) Unconscious processes result in a random fusion of memory structures. This is blind and undirected but selected by the tendency to retain the more appropriate attributes and fusions which contribute to the solution.

One reason why this approach does not enjoy much currency today in cognitive psychology is that, although providing a useful conceptual framework for the discussion of problem solving and the carrying out of creative or original works, it does little to tell us how these are done. Current thinking on human cognition tends to view thought processes in terms of information processing theory which will, in outline form, be familiar to anyone who has a rudimentary knowledge of computer science (Posner 1973).

The whole field of 'design studies' which is even now a keenly contested field of debate has, like any relatively young discipline, been characterised by large numbers of 'theories' and 'conceptual frameworks' and relatively little empirical evidence. This of course increases the difficulty for an observer in identifying the main characteristics of the design process. A consideration of differing definitions is often helpful. The literature surveyed produced almost as many definitions of design as there were writers on the subject. A diverse selection follows:

> 'To initiate change in man-made things'
>
> (Jones 1970: 4)

> 'Finding the right physical components of a physical structure'
>
> (Alexander 1963: 83)

> 'The optimum solution to the sum of the true needs of a particular set of circumstances'
>
> (Matchett 1968: 166)

> 'The architecture of Beaubourg becomes an expression of the process of building; the optimisation of every single element, its system of manufacture, storage, transportation, erection and maintenance all within a clearly defined and rational framework'
>
> (Rogers 1980: 112)

It is accepted that the quotations from Richard Rogers and Christopher Alexander is not strictly a definition of design but it is relevant all the same. The differences among these definitions are indicative of the complexity of the issues involved. The words of Jones and of Matchett are clearly intended to refer to design in the general sense. They are presented here though, for the light they can shed upon building design, the form of design which is the concern of this chapter. The definition of Jones is difficult to argue with but it seems to be based on the doctrine of reductionism carried to its logical conclusion and tells us little. By the same token, a biochemist could define design activity in terms of the chemical changes taking place in the neurones, although testable this would not be of assistance in this context.

The definitions of Matchett and Rogers are appealing but they assume that the true needs of the particular set of circumstances *can* be established and furthermore that the *optimum* can be located. However, they are helpful, in that they hint at one of the central difficulties of design, which will be referred to again below, namely that in design the task of identifying problems, or 'problem finding', may be as great as or greater than, the task of 'problem solving'.

However, there is something missing from all of the above definitions. Surely design is more than a mere exercise in optimisation. The definition of Norman Foster embodies the principles of:

(1) Optimising many variables to satisfy needs.
(2) The whole, being more than a mere assembly.
(3) That design is not a process at all.

> 'Design is really a tool. It is a means of integrating and resolving the inevitable conflicts that range from public/private to socially acceptable/commercially viable, in order to reconcile the artistic aspects of making a building with cost, time and quality control. By trying to optimise all the givens within a consistent framework of values upon which design decisions are based, we try to arrive at a whole which is more than the sum of its parts.'
>
> Foster (1980: 138)

A Three-way View of Design

In his influential text first published in 1970, Christopher Jones presented an analytic, although conceptual, view of design in the context of his work on design methods. He proposed that while working on a problem the designer was engaged simultaneously in three types of activity.

(1) Creativity; or the designer as a black box.
(2) Rationality; or the designer as a glass box.
(3) Control over the design process; or the designer as a self-organising system.

Clearly they are only separable for the purpose of discussion but they do seem to sum up rather well the type of thought processes required.

Designer as a Black Box

The main characteristics of black box design are:

(1) The output of the designer is governed not only by the inputs received from the problem in hand, but also from past problems and other problems. Each new task is viewed in the 'light of experience'.

(2) The capacity to produce outputs relevant to the problem depends on being given time to assimilate and manipulate within himself or herself images representing the structure of the problem as a whole. During a long search for a solution he or she may perceive a new simpler way of structuring the problem – the 'leap of insight'.
(3) Intelligent control over the form in which the problem is fed into the human black box is likely to increase the change of obtaining relevant output.

Designer as a Glass Box

Most of the formalised design methods produced in the 1960s and 1970s for the design of buildings come under this heading. They tend to envisage the designer as a human computer acting only on the information that is entered, and then following through a planned sequence of evaluating synthetic and analytical steps and cycles until he or she recognises the optimum solution.

The more common characteristics of the glass box methods are:

- Objectives, variables and criteria are fixed in advance.
- Analysis is completed or at least attempted before solutions are sought.
- Evaluation is largely logical.
- Strategies are fixed in advance.

Designer as a Self-Organising System

As the designer works at the problem, various avenues will be explored as possible sources of solution. There are far too many for each to be fully evaluated, so as work continues on the central task. The designer needs to constantly enquire of himself or herself whether this route is likely to prove fruitful or not. In fact there appear to be two choices:

- Make a black box (arbitrary) choice of routes to be explored.
- Plod away at the impossible task of evaluating each proposal separately.

In reality it seems that designers often take neither of these two, but work on their problem by dividing the available design effort into two:

- That which *carries out* the search for a suitable design.
- That which *controls and evaluates* the pattern of search (strategy control).

By doing this it is possible to replace blind searching through alternatives with an intelligent search that uses both external criteria and the results of partial search to find short cuts across unknown territory. Strategy control seeks to relate the results of small pieces of search to the ultimate objectives even if these are in a state of flux.

The above view of the activities of the designer is based on Jones' (1970) approach to design methods. The dearth of empirical work means that there are as many views as writers and little means to judge rationally which are the more valid. It has been presented here to add to the 'character sketch' of the design process which is being built up in this chapter.

Design Methods

The 1960s and early 1970s produced a number of 'design methods', formalised methodical approaches to design. None of these methods found regular use among building designers. It is of course debatable whether there can be such an entity as a 'design method'. This is referred to again later. However inappropriate the methods were for practical use, they did reveal something about the design process. The earlier methods were very mechanistic and rational. In the 1970s the debate became softer with increased emphasis on user participation in the design process. There is little doubt though that the two methods which most of all captured the public imagination were those of Alexander (1963, 1964) and of Jones (1970). They are worth considering briefly.

Alexander's Method

Based entirely on rationality, the essence of the method was as follows:

> 'The form is the solution to the problem. The context defines the problem. The ultimate object of design is form. We need to fit the form to its context.'
>
> Alexander (1964)

Alexander saw the process of achieving fit between form and context as a negative process of neutralising the incongruencies or forces which carried misfit. Given that there are limits to a person's capacity for mental arithmetic bigger problems need to be set down on paper in a logical way. Set theory was used as an analytical tool. The designer attempted to 'organise' his or her view of the problem by 'decomposing' the problem into a tree-like hierarchy of its subsets. For a real problem the hierarchy would be very large with many hundreds of elements. The aim was to identify the discrete sub-problems of decomposition. Independent solution of the sub-problems so identified would result in a solution to the design problem.

Although mathematically elegant, the method proved to be unworkable in practical terms. It did however make an important contribution to our knowledge of the design process. It took the view that the design process was a highly complex tree-like hierarchy where each compṗnent interacted with many others. Shortly afterward Alexander rejected his method completely (Alexander 1966, 1971) and moved on to a user orientated approach where the designer uses his or her knowledge to assist the user to design his or her own building (Alexander *et al.* 1975, 1977, 1979).

Jones' Method of Systematic Design

This was an attempt to represent logically the overall design process. It was not specifically directed at architectural design. The main stages are summarised below (Jones 1970):

(1) Analysis
  (a) Random list of factors.
  (b) Classification of factors
  (c) Sources of information.
  (d) Interactions between factors.
  (e) Performance specifications.
  (f) Obtaining agreement.
(2) Synthesis
  (a) Creative thinking.
  (b) Partial solution.
  (c) Limits.
  (d) Combined solutions.
  (e) Solution plotting.

(3) Evaluation
  (a) Methods of evaluation.
  (b) Evaluation for operation for manufacture for sales.

The analysis stage is a divergent process where the problem is explored and a list made of all the relevant factors. This listing and classification of factors is intended to assist in the definition and organisation of the problems and sub-problems to be solved. Performance specifications are written in order to separate the problem from the solution. The requirements and factors are rewritten as performance specifications with no reference whatever to shape, materials and design.

The synthesis stage is one of convergence. It is the black box stage in this design method. The method aimed not at finding a single solution, but at establishing a range of solutions and clarifying the points where they fit or do not fit the specification. Evaluation was taken to mean any method by which deficiencies in the solution chosen may be detected before it becomes prohibitively expensive to correct them.

The method was presented as a means of resolving the conflict between logical analysis and creative thought. Jones attempted to keep logic and imagination apart by external means, i.e. by keeping an external written record of all the ideas at various stages while at the same time allowing the mind freedom to produce ideas, solution hunches, etc., without confusing the process of analysis. The method does give a mechanistic representation of the process of design, but clearly it is not the kind of method which many designers would find it pleasing to work through in detail because of its sequential and severely logical nature.

However, the contribution made by Jones' method is that it concentrates attention on what has become known as the 'analysis synthesis appraisal loop'. This loop, which was first proposed by the Building Performance Research Unit and later developed by Maver, is one of the few views of design upon which most writers agree (Markus *et al.*, 1972, Maver 1977). It has underpinned much research and development in computer aided architectural design since then. It is summarised in Fig. 10.1. Neither Jones nor Alexander gave, at that time, any formal recognition to the iterative nature of the design process. The analysis synthesis appraisal loop makes up for the shortcoming.

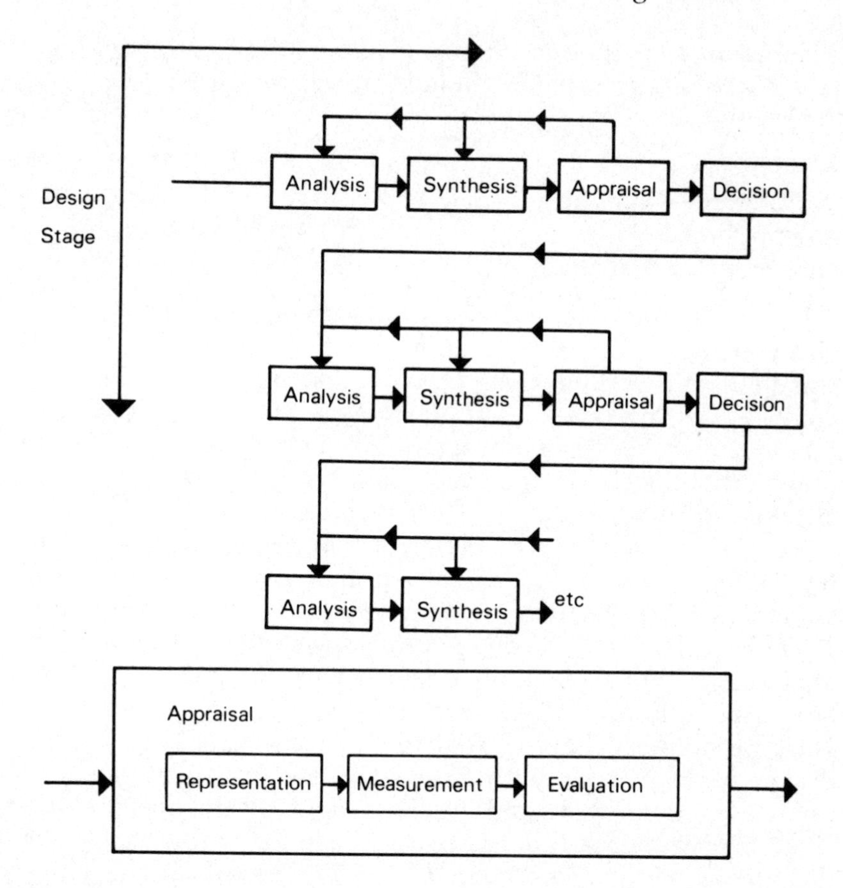

Fig. 10.1 Analysis synthesis appraisal loop. (*Source*: Maver (1970))

Empirical Work

There have been relatively few reports of experimental studies of the design process and the behaviour of designers. An excellent paper by Lera (1983) reviewed ten such studies. Some authoritative and diverse empirical work of value has been reported since then, in the journal of the Design Research Society and elsewhere but this later work is far less easy to apply in the domain of interest of this chapter. Bucciarelli (1988), Mackinder and Marvin (1982), Newland *et al*. (1987) and Schon (1988). Two of the studies reviewed by Lera are worth pursuing briefly.

Eastman (1970), working at Carnegie Mellon University, was one of the first to study the work of designers under controlled experimental conditions. He considered the forms of representation which were used by building designers, i.e.:

- Words.
- Numbers.
- Plans.
- Flowcharts.
- Sections and perspectives.

He concluded that there was a significant correspondence between the types of constraints which were identified by designers when working on a problem and the form of representation which they used. The most often quoted example being, that in the design of a bathroom it was only when a section was drawn, that the ability of a child's hand to reach the taps became an issue to be solved in the design solution.

Simmonds, of Oxford Polytechnic in the United Kingdom, reported in 1980 on work carried out with twelve graduate students of Architecture in an American University (Simmonds 1980). He considered the decision-making strategy of the students at three different levels, only one of which (the overall level) will be discussed here. At the overall level he found that there was a wide variation in the way that the subjects went about solving design problems. Some analysed the problem, generated solutions and then considered implementation of the design. Other subjects generated solutions, used the solutions to derive problem definitions which were then tested against the brief. Others began by considering the resources available and the constraints on their use.

From these results we may derive the following general conclusions with respect to the building design process and the behaviour of designers.

(1) The number of solutions evaluated varies from designer to designer.
(2) The nature of the solution eventually chosen is related to the representational technique used in carrying out the design problem.
(3) Designers use a variety of representational techniques and also use different logical sequences in carrying out the design.
(4) If the above are true, there must be more than one 'correct' solution to each design problem. The correctness of the answer depends on the way the problem was solved. There

is an analogy with mathematical problems using, say, mental arithmetic, paper and pencil and computer, where increasing degrees of accuracy are required for the answer to qualify for being called 'correct'.

(5) Conclusion number (2) above has major implications for computer aided design (CAD) and computer aided architectural design (CAAD). The increasing use of software based representational techniques by architects may change the type of solution reached to certain problems. Work needs to be done to check if the change is always for the better. Can CAAD produce solutions which were previously impossible?

Synthesis: Building Design

The discussion above has revealed many of the salient features of building design, the exploration or 'problem finding' nature of the early stages, followed by a synthesis and a proposed solution. The iterative nature of design has been mentioned but historically this problem has been dealt with inadequately. An example of this iteration is where a definite decision is made regarding the type of lights and light fittings, say, for an office accommodation, relatively late in the design. The heat gain resulting from this decision could be outside the range assumed by the services designer, who may then find that the air-conditioning or heating plant may need to be adjusted to take this into account.

The truth may be that it is not possible to deal with this type of interaction exhaustively and in a methodical way. In order to allow for all cases, the minimum number of decisions to be anticipated would be the total number of combinations of all of the decisions made separately. The number of possible solutions becomes very large indeed and the effect of each could not be appraised without resort to very powerful computer aids. Iteration is necessary as the solutions to certain sub-problems create new problems and alter the feasible space for other sub-problems, some of which may already have been 'solved'.

The later work of Alexander (1975, 1977, 1979) and proposals made by Krauss (1970) and Gill (1980) attempt to deal with the problems of non-linearity and the interactive nature of the decisions. The method as observed by Krauss was a dialectic or 'to'ing and fro'ing' between the design problems and their solutions with continual iteration and adjustment of previous solutions. With each iteration more was learned about the nature of the problems. Gill suggests an 'adoptive approach' which is in conflict with those who have sought to derive a design method but does serve at a general level as a description of the design process.

> 'A decision process which has no long-term action other than heading off undesirable trends is more likely to achieve good results than a process which seeks to steer society in a predetermined direction, selected on inadequate grounds'.
>
> (Gill 1980)

It should be recognised that very few designers will carry out the task in the same way and even if they did, it might not be possible to show that this was indeed the case. It is for this reason that the descriptions of the design process which have endured have tended to be at a general and procedural level and not at a detailed level. Examples of this kind are the *RIBA Plan of Work* (RIBA 1965) and the descriptions of Maver (1970, 1977) and Jones (1970). The *RIBA Plan of Work* has the major fault of not acknowledging the iterative nature of the process, but otherwise it provides a very general description of the movement towards the detailed design solution. The approach based on the analysis synthesis appraisal loop takes account of all the problems raised above at a general level.

An often asked question is 'is building design (or design generally) an art or a science?' The design methodologists of the 1960s and 1970s seemed to suggest the latter. It has been argued that the idea of rational design as it is understood, ie conforming to the 'orthodox area of science and deductive logic' is not appropriate or helpful (Abel 1980). March (1976) has criticised the 'Popperian models' as 'pernicious'.

> 'Logic has interests in abstract forms. Science investigates extant forms. Design initiates novel forms'.
>
> (March 1976)

Indeed, it has been shown that the concept of 'scientific method' itself is in a state of 'epistemological chaos' (Cross *et al.*, 1981). Furthermore, Cross, Naughton and Walker of the Open University, the same authors, present an attractive view of design (when applied to building design) namely that, design is a technological activity identified by these fundamentals:

(1) Practical tasks. Technology is orientated not towards understanding but towards actions or solutions to defined problems.
(2) Different kinds of organised knowledge are used, i.e:
- Scientific knowledge.
- Craft knowledge.
- Design knowledge.
- Organisational knowledge.
- Managerial knowledge.

(3) The activity takes place in an organisational setting.

Thus, three decades of research into the design process have produced, it must be said, disappointing results. Notions of 'method' and logic in design have not been actually discredited but rather refocused into the background as a relatively minor attribute of the process. Creation of something new is, in the final analysis, a mysterious, personal activity. Logic plays a part, as do previous experience, images, wishes and hunches. The most authoritative works on the nature of design lift the field to a new plane where the systems ideas of the 1960s and 1970s play only a minor part. An excellent overview of this mature approach may be had from Canter (1977, 1984), Lawson (1980) and, most comprehensively, Zeisel (1984).

In conclusion, the various issues discussed in this chapter may be drawn together.

The dominant characteristics of building design are:

(1) It is very complex and possesses a large solution space.
(2) The problem to be solved is ill-defined.
(3) The process is iterative due to the large amount of interdependency among decisions.
(4) The solution reached depends on the techniques used and the way in which the designer approached the problem. These vary from designer to designer, thus it may be said that, to a degree, design is personal.

Chapter 11

Forecasting Prices and Costs

Introduction

In this final chapter we will develop an approach to the assessment of the models used in forecasting construction costs and prices. We will also demonstrate, in passing, that forecasting for purposes either of capital investment planning or control in construction, has developed almost entirely independently of advances in business and economic forecasting, To these ends, we will first consider the notion of 'accuracy' in forecasting. Second, we will review the principal developments in business forecasting. Third, we will focus on the field of construction by reviewing the development of cost and price forecasting techniques. Fourth, we will introduce a conceptual framework for the assessment of these models. Fifth, we will present three typical models from this field. Sixth, we will assess the robustness of these models by using the framework developed earlier. Finally, we will attempt to draw some conclusions about the current state and future potential of construction price forecasting techniques.

Accuracy of Construction Forecasts

How good are the models of building costs and prices commonly used in practice? One might legitimately respond that the question is ill-defined and that a more precise formulation would seek to establish how 'accurate' the models were. The common response to a question like this is to obtain a large sample of building cost estimates made by, say, quantity surveyors, using some form of modelling technique, such as elemental cost planning, and to compare these with the actual prices given by the successful tenderers. This approach while undoubtedly important (it goes directly to the bottom line) is, it is suggested, of limited usefulness. Of itself, it supplies a result but gives no understanding of why or how this particular result was achieved. The accuracy of the model's output has been measured, not the accuracy of the model.

For professional purposes, in the short term, maybe being right is all that matters. If however, we want to continue to produce good forecasts when conditions change, then some element of causality is necessary. A consideration of the etymology of the word 'accuracy' itself (exact, correct, from the latin '*accuratus*', meaning performed with care) suggests that one should not just consider the output of the model but the performance of the model generally. While the accuracy of building price predictions can be measured in a word or a number, a more detailed appraisal of model performance is necessary in order to effect an improvement in the accuracy of the output. This approach will be exemplified below by a consideration of the 'accuracy' of some of the more common methods of modelling building cost and price.

There is, it must be said, some dispute about what actually is being forecast. Is it the lowest tender? What if the lowest tender is a mistake? What if the lowest tender is not a mistake but, in any event, it turns out to be a wrong estimate of what it eventually turns out to cost the contractor and thus soaks up its prior calculation of profit? Perhaps we are trying to forecast the average of all the tenders received. This average will be influenced by errors at both ends of the range and 'cover prices' at the top end of the range. Additionally, the actual cost of constructing the project will not be known until the project is complete and all accounts have been agreed, if not settled.

Let us assume that in most cases the lowest tender is not an error. In some cases the lowest tender will be an error and this will be obvious to the building cost consultant who may recommend acceptance of another tender. What we are trying to forecast is, therefore, the *accepted* tender. This accepted tender will represent the winning contractor's *estimate* of the market price for constructing the project. The estimate may or may not be a correct estimate of the outturn cost (plus profit) of the project. This highlights the difference between forecasting construction tender prices and business forecasting. The latter is concerned with forecasting phenomena which may subsequently by measured. Instances of this could include, say, forecasts of sales and inflation. In building economics we are mostly involved in forecasting an estimate.

We need to clarify further what is meant by 'accuracy'. There is a common tendency to believe that the more precise a statement or figure is the more accurate it is. For example the psychological weight attached to a computer print-out which gives the answer to a problem in six decimal places. 'Accuracy' may be used to denote 'correctness' or care in performance as suggested above. Kamarck (1983: 2) provides a useful illustration:

> 'On Cape Cod, where the pace of life is unhurried and casual you may ask a craftsman in June when he will come and repair your fence. If he answers "Sometime in the Autumn" he is being accurate but not precise. If he answers "10.00 am October 2nd" he is being precise but not accurate – it is almost certain that on October 2nd the fish will be running and he will be out in his boat. One of the recurring themes that you will find in our discussion is that too often in economics the choice is between being roughly accurate or precisely wrong.'
>
> (Kamarck 1983: 2)

Brandon and Newton (1986), in a description of knowledge based approaches to building cost modelling, state that in order to be effective and to achieve practical implementation the mathematical models need to allow human judgement to be exercised over the processes. Bowen and Edwards (1985) suggest that rather than just trying to improve current practice, researchers need to tackle more fundamental issues designed to bring about a change in 'thinking'. Raftery (1984) suggests that insufficient attention has been devoted to the limitations of cost models. Decision makers find it difficult to evaluate the performance of models, because they lack a consistent framework with which it can be measured. Taking a systems view of the problem, such a framework was proposed and the criteria discussed.

This chapter attempts to apply this conceptual framework for the assessment of model performance in the context of recent developments in the field. It will show that human judgement may be exercised in a systematic way over so called 'black box' models. There will be no startling results or complex quantitative evaluations. The intention is merely to demonstrate, in a simple way, a method of organising one's thinking in order to apply common sense judgements to the strengths and weaknesses of models.

Business Forecasting

Techniques of business forecasting are commonly divided into three categories. Namely, qualitative, time series and causal methods. (Barron and Target 1985) Qualitative methods are based on judgement. They are distinguished from mere guesses by their systematic nature. One of the most well-known qualitative methods is the *Delphi Peer Group Forecast*. The method is named after the town in ancient Greece wherein resided the famous oracle. The method works as follows: a group of subject experts is identified. The members of the group are kept physically separate. The chair asks each member of the group to make a forecast, over the relevant time period, of the variable under investigation. The chair receives and summarises the forecasts. The summary is communicated to each group member. Group members are then asked to amend their forecasts in the light of information in the summary. The new forecasts are summarised and this is communicated to the members. The process continues until either there is a consensus or the group members no longer wish to amend their forecasts. In either case, the final result is the *Delphi* forecast. The physical separation of members is designed to nullify the effects of strong personalities and other undesirable features of group interaction.

Time series methods involve predicting future values of a variable from observations of its historical behaviour. In a sense, time series methods use 'mathematical' extrapolation to forecast future values of a variable taking into account, trend, seasonality and cycles. Detailed experiments on the use of time series methods for the forecasting of building cost indices have been reported by Fellows (1988). Being computer based, they are cheap and quick to use. They are best used when quick, short term forecasts are required in relatively stable conditions. They are not good at forecasting when there are changes in underlying conditions which may cause significant new trends or indeed turning points. Intuitively, time series methods do not seem appropriate in a dynamic and volatile economic environment.

Causal methods, as the name implies, involves the identification of some leading variable which has a causal relationship with the variable under investigation. A model is built using regression or correlation. For instance, the internal volume of a building has a causal relationship with the size of the cooling plant necessary for an air conditioning system. A simple model could be built by taking building volume as the independent variable and cooling load as the dependent variable. If data were then collected for a number of buildings it would be a simple step to fit a line or curve to the scattergram. Internal volume is not the only variable which affects the size of the air-conditioning plant. A more complex model could be built using, say, volume, external climatic conditions and internal heat gain as independent variables. Causal models are used in business forecasting when there is good reason to believe that conditions are unstable.

Business forecasting is usually carried out in the context of a forecasting or management information *system*. At its simplest this will include the following (Barron and Target, 1985):

(1) An assessment of management's information needs, i.e., what information is needed and when, in order to manage the process in question?
(2) A specification for the forecasts that are necessary, given the information needs above. This should include the required level of accuracy, time horizon and the frequency with which forecasts will be needed.
(3) A conceptual model for the forecasting system. This should identify all the key factors which have an influence on the variable to be forecast. This descriptive model should identify whether there are any historical patterns, or changes of underlying conditions. From the specification and the conceptual model, it should be possible to identify the type of forecasting method most suited to the problem.
(4) A survey of data, its sources and its quality. This should also identify data which is not available.
(5) A method of making the relevant forecast.
(6) Some testing or validation of the forecast.
(7) A Systematic method of allowing human judgement to be incorporated in the forecast. Quantitative and, to a lesser extent, causal techniques tend to assume that the conditions of the past will continue. Methods which do not allow human intervention may lose credibility in turbulent periods and thus fall into disuse. On the other hand, the method of incorporating judgement should be such that personnel are held accountable for any interventions they do make.
(8) A system of monitoring the performance of the entire forecasting system. Statistical tests of the results under varying sets of circumstances should be made periodically while the information system is in operation.

It is interesting to note that the *technique* of forecasting, the 'number crunching', is only one stage out of eight listed above.

Developments in Building Cost and Price Modelling

First Generation Models

In an attempt to organise our thinking we can consider developments as having taken place in three distinct generations. The first generation which began in the late 1950s and continued up to the late 1960s was characterised in the United Kingdom by a procedural (elemental cost planning) approach and in the USA by much published work on bidding models.

In the United Kingdom, the *Building Cost Information Service* (BCIS) was set up in 1962. The original *raison d'être* was to build up a data bank of building cost analyses which would be submitted by members and then made available to all other members. These cost analyses, suitably updated for time quantity and quality, formed the basis for elemental cost plans which were used to derive square foot prices which formed the basis of design cost control and tender price prediction. The 'updating' for time, quality and quantity was and is frequently carried out in a mathematically naive way using straight-line adjustments for differences in floor area, building cost and tender price indices for adjustments in time, and 'professional judgement' for quality adjustments. As more design information became available, approximate quantities could be measured and quality adjustments could be made more accurately and more objectively.

In the USA, at this time, the published work was in bidding models. The long running bidding model controversy between Friedman and Gates in the Journal the American Society of Civil Engineers started in 1956 and ran right up to the early 1980s (Friedman 1956, Gates 1967, Rosenshine 1972, Dixie, 1974, Benjamin and Meador 1979).

Second Generation Models

The second generation began around the mid-1970s, this generation of cost modelling and forecasting was characterised by intensive use of regression analysis on both sides of the Atlantic (Kouskoulas and Koehn 1974, McCaffer 1975, Buchanan 1973,

Gould 1970, Wiles 1976, Bowen 1982a, 1982b, 1984). Second generation models appeared to increase and multiply in a direct relationship with the growth in availability and reduction in cost of micro-computers. The models were frequently derived from statistical packages operating on mainframes. Once the equation parameters had been defined these were then mounted on desktop micros as rapid response models.

Third Generation Models

The third generation approach to modelling appears to have begun in the early 1980s. This third generation has two central characteristics. Firstly a willingness to admit to the existence of uncertainty and imprecision and a desire to take account of this by carrying out probabilistic estimates frequently based on Monte Carlo techniques (Wilson 1982, Raftery 1985a, 1985b). As with second generation models, the driving force behind these approaches was probably the increasing availability of fast computing facilities at lower prices. Additionally the move to probabilistic approaches was a late response to changes in thinking in science and technology generally in the 20th Century. The second characteristic of third generation models is also computer linked, it is the current interest in artificial intelligence and knowledge based computer systems (Brandon and Newton 1986). The knowledge based aspects of third generation models mark an interesting return to the first generation approaches in the United Kingdom where great reliance was placed on professional skill and judgement. This time round however, researchers are attempting by knowledge elucidation to 'codify and organise professional judgement'. This again reflects a broad trend in science and technology generally in the 1980s.

The three generations of building cost/price models may be seen as a reflection of available techniques, computational aids and the contemporary mind-set with which problems are approached. Publications in the third generation era have sometimes been accompanied by exhortations to researchers to build up a 'scientific base' for what was previously a technical subject. Bowen and Edwards (1985) have, in posing the apposite question, 'who needs a new paradigm', pointed out that the demand for a move to a more 'scientific base' appears to come mainly from academics and not from practitioners. It may not then be too unkind to suggest that eager to publish, academics, given the generational conditions mentioned above, sometimes produce exercises in fashionable techniques and thinking. Popper (1964) and more recently Kamarck (1983) have warned against this drift towards scolasticism, 'apparently an almost irresistible temptation in academia in all subjects' (Kamarck 1983: 122). Therefore, while attempting to increase the rigour of the subject, we would do well to be ever mindful of the purpose of these models, i.e. solving the real world problems of predicting building price and exploring in some detail the various economic implications of varying design strategies.

A Conceptual Framework for the Assessment of Model Performance

Consider the real world context of the economic models under discussion here. The building designer is in the position of juggling many decisions at the same time and of attempting to optimise the overall result of decisions affecting all of the design subsystems, for instance, geometry of internal spaces, frame and services. Some models are designed to consider, in detail, subsystem problems such as air conditioning, plant, or lift design. Others seek to predict the tender price expected for the whole building. Clearly, very different model characteristics will be required for these tasks but there are some features which apply regardless of this. This is the level at which the conceptual framework illustrated in Fig. 11.1 seeks to operate. The model may be viewed as part of a chain which leads from raw data through some kind of model and output on to a decision-maker, in this case the design team.

If we discount the internal cognitive processes of human decision making, four criteria useful for the assessment of performance remain. These are clearly based on a systems view of the problem and are as follows:

- Data.
- Data/Model interface.
- Model technique.
- Interpretation of output.

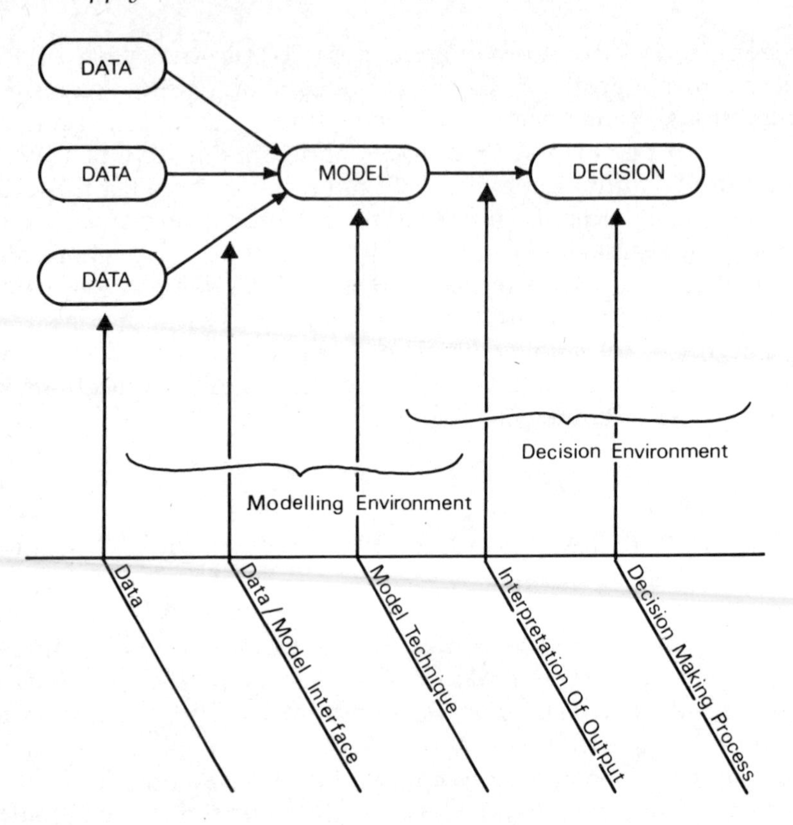

Fig. 11.1 A conceptual framework for the assessment of model performance

Three Models

Model 1 – An Element Based Floor Area Model

The Building Cost Information Service (BCIS) which was set up in 1962 supplies a data bank of analyses of building costs which break down the accepted tender into various elemental categories such as substructure, finishes, services etc. The technique of cost planning using this data base is described in some detail by Ferry and Brandon (1991).

Consider here a form of representation, the preliminary cost plan may be defined as:

$$C = \sum_{E=1}^{6} (t.q.qu.R)E \qquad (1)$$

where E = the main element headings. (Substructure, superstructure, internal finished, fittings, services, external works.)

t = the time adjustment, not all elements will suit the same index, for example the services element will often by updated by a mechanical and electrical services price index and the other elements by a tender price index for building work which excludes these services.

q = the quality adjustment which attempts to account for differences in the level of specification of the element in the building chosen for comparison and that of the project in hand.

qu = the quantity adjustment. This is usually assumed to be linear, by merely substituting the new building area for the old one and not applying any adjustment for economy or diseconomy of scale.

R = either the element unit rate or the cost of the element expressed per unit gross floor area.

Consider now the detailed form of the model:

$$C = \sum_{E=1}^{32} (t.q.qu.R)E \tag{2}$$

Clearly there is much more information at this detailed level now including 32 elements. Each of the six elements is broken down further, for example 'superstructure' becomes, frame, upper floors, roof, stairs, external walls, windows and external doors, internal walls and partitions, internal doors. However this extra detail is not supported by a refinement of the functional relationship. This in fact remains the same as each subheading is expressed in terms of unit floor area. The method of derivation may be different, at this stage approximate quantities may have been computed. This may be expressed as follows:

$$C = \sum_{F=1}^{N} (Q.r) + P + K \tag{3}$$

where F = the number of items of finished work, measured separately.

Q = the quantity of any of the items in F. The unit depends on the item (examples are xm^2 of internal partition, Y no. of internal doors, ZM of drainage trench 1.00 m deep, etc.)

r = the unit rate for each item of finished work F which includes usually for the materials, labour and overheads necessary to fix the work in place.

P = preliminaries, indirect costs as described above.

K = an amount added to or taken from the total to take into account market considerations etc.

Model 2 – A Regression Model

Consider as an example here the pre-design cost estimation function presented by Kouskoulas and Koehn (1974). This was based on a sample of 38 buildings which was widely based and appeared to be a random sample rather than one designed to take into account the range of values of variables.

The function proposed was:

$$C = -81.49 + 23.93V_1 + 10.97V_2 + 6.23V_3 + 0.167V_4 + 5.26V_5 + 30.9V_6 \qquad (4)$$

where C = some cost measure of buildings (in the derived model C became the dollar cost per unit floor area).

V_1 = locality index; based on cost of living and wage differentials and taken from a publication on building construction cost data.

V_2 = price index; compiled in a similar way to V_1 from municipal statistics.

V_3 = building type; an index compiled from the cost differentials among the various building types.

V_4 = height index; measured by the number of storeys.

V_5 = quality index; this attempts to measure: (1) the quality of workmanship and materials used in the construction process, (2) the building use, (3) the design effort, (4) the material type and quality used in the components. V_5 was derived from the expression:

$$V_5 = \frac{1}{k} \cdot \sum_{i=1}^{K} I_i.C_i \qquad (5)$$

where k = the number of building components.

C = the portion of cost of the ith component.

I = an integer between 1 and 4 (corresponding to fair, average, good and excellent) arbitrarily applied to that component.

In the derived model the number of building components k was 8, these were the building use (multi-tenancy, single tenant, mixed, etc.), building design (minimum, average, high loads etc.), exterior wall, plumbing, flooring, electrical, HEVAC, elevator.

V_6 = technology index; this attempted to take into account extra costs of special types of buildings, or the labour and material savings resulting from the use of new techniques. $V_6 = 1$ for normal situations, $V_6 > 1$ for extra costs (i.e. chemistry lab = 1.45, bank = 1.75) and $V_6 < 1$ for savings as a result of technology.

Indices for these six independent variables were applied to the sample of buildings, a least-squares analysis was carried out on each of the linear equations which, when solved simultaneously, gave the function shown above.

Model 3 – A Probabilistic Model

Here we will consider a probabilistic approach to the estimation of net construction costs (not prices) by building contractors presented by Raftery (1985a, 1985b). 'Model' in this context may well be a misnomer, the approach is a simple Monte Carlo evaluation of a construction cost estimate compiled using the estimators subjective perceptions of probability distributions for each of a set of reasonably independent sub-systems.

Construction risks are frequently project specific. These are sometimes accounted for by estimators by adding a risk premium to the tender cost estimate. Consider the tender summarised in Table 11.3. This was prepared by an estimating group in a workshop on estimating techniques. The estimate is for an office building in Damman, costs are given in Finnish marks.

Estimators have long been aware of constructions risks but traditional methods of including them in the calculations have tended to obscure the issues. The approach described here attempts to capture the estimator's perception of risk in a realistic way by eliminating the need for the estimator to make one best estimate for each variable. Instead he or she may enter into the calculation a description of his or her complete judgement about the variable in question. These judgements are made in the form of probability distributions. The method consists of combining these probabilities and calculating the resultant. This is done in a non-mathematical way by generating a large number of projects with the general characteristics of the one in hand and observing the pattern of the results. In fact, a statistical analysis is made of the results just as if they were a sample of actual projects. This method of analysing risks for large capital investments has been described in detail by Hertz (1979).

Consider the estimate in Table 11.1. If the percentage additions and mark-up are removed the net cost estimate comes to 76.505 million mark as in Table 11.2. Each figure in the right hand column represents the estimating groups' 'most likely' costs. The estimators then went on to produce another estimate summary in which they attempted to portray the degree and shape of uncertainty which attached to each of the variables. This is illustrated in Table 11.3.

Table 11.1 Deterministic estimate: Case study building, Dammam office.

Direct Costs		FIM/m^2
Materials		1800
Labour		600
Mechanical Sub-contract		800
Electrical Sub-contract		500
Indirect Costs		
Plant		900
Staff		900
		5500
Escalation 7%		5885
Mark-up		
Agents		157
Risk 5%		392
Profit 15%		1177
Insurances bond 3%		235
	1300 m^2	7846
		1.02Mmk

Table 11.2 Dammam office, net cost estimate.

	FIM/m^2
Direct Cost	
Materials	1800
Labour	600
Mechanical sub-contract	800
Electrical sub-contract	500
Indirect Costs	900
Plant	900
Staff	5500
Escalation	385
	5885
1300 m^2 × 5885 = 76 505 000	

Table 11.3 Simple Probabilistic Estimate.

	Min	*Most likely*	*Max*
Direct Costs			
Materials	1700	1800	1850
Labour	550	600	700
Mechanical Sub-contract	800	800	900
Electrical Sub-contract	500	500	600
Indirect Costs			
Plant	850	900	900
Staff	850	900	1000
Escalation	6%	—	8%

The Monte Carlo approach merely involves entering the estimate in this form into a short computer program which then proceeds to 'build' the building many times. On each pass through the project the computer selects a cost for each item which is chosen from the input distribution for that item. This input distribution represents, as we have seen, the estimators perception of the feasible range for that item and the relative probability of the values within that range. In this way we are able to observe the effect of the combined probabilities. Results are usually presented as frequency histograms and cumulative probability graphs. In the example here the estimate was simulated 500 times and it was shown that the deterministic estimate

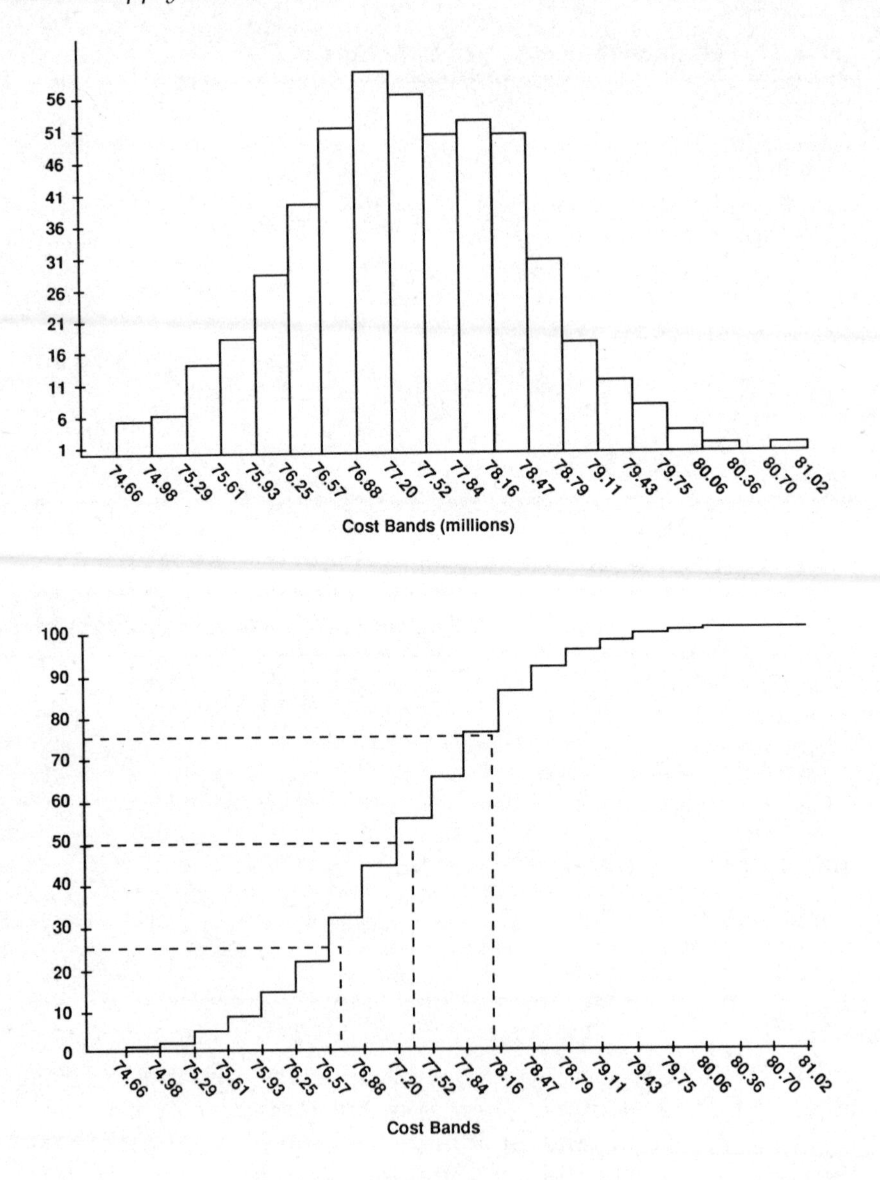

Fig. 11.2 Probabilistic estimate

of 76.505 million marks has more than a 75% chance of being exceeded, assuming that the input distributions were accurately defined. Using this approach it is possible to attach some sort of numerical evaluation to the degree of risk contained in the estimate. This is illustrated in Fig. 11.2.

This is a simplistic overview of Monte Carlo evaluation. More detailed discussions are available in Hertz (1979), Hertz and Thomas (1983), Pouliquen (1970) and Reutlinger (1970). An alternative approach to a similar problem has been presented by Turunen (1984).

Analysis

The models above are presented only in summary form. These models may now be considered in the context of the systems criteria identified in Fig. 11.1. The criteria will be considered in turn across each of the three models. The discussion is summarised in the evaluation matrix presented in Table 11.4.

Data

The data used in the elemental floor area model is conventionally derived from elemental cost analyses held at the BCIS. These cost analyses are compiled by quantity surveyors from the priced bills of quantities of the successful tenderer for a particular project. The tender price for the project is broken down into the unit rates for particular items of measured work.

The process of cost generation, recording and the eventual derivation of what is euphemistically called 'data' is riddled with inconsistencies. Subjectivity creeps into the data under discussion here in two ways. Firstly, during the recording of the event and secondly during subsequent transformations of the

Table 11.4 Evaluation Matrix: Identifying Key Problems

	Floor area	*Least squares*	*Monte Carlo*
Criterion/Model	1	2	3
Data	Transformations	Sample bias	Data collection
Data/Model Interface	Good fit in early stages – less well later	Global data/ global model	Good fit
Model Technique	Linearity	Linearity effect of sample	Independence problem
Output	Inflexible	Inflexible	Interpretation

data to produce 'information' for various procedural requirements. Records for the time taken to perform various tasks in a project may be distorted, for example by bonus schemes which apply to some tasks and not to others.

The unit rate for finished work contained in a fully priced bill of quantities are highly unlikely to be robust price data. The successful contractor has offered to deliver the project for an agreed sum. Any subsequent breakdown of the figure into units of finished work will at the very least be notional. Further, a rational firm would seek to use the breakdown to optimise its cashflow, say, by overpricing work items in the early stages of a project.

Two major transformations of data occur. The first is when the sum of the resource costs is spread over the unit rates to produce the priced bill of quantities with the attendant loading and tactical decisions. The use of labour only sub-contractors renders unit rates less appropriate. The second is when the unit rates are subdivided and clustered into element costs. These processes are illustrated in Fig. 11.3. To complete the picture of the cycle of use of this data it should be added that the elemental rates tend to be used by quantity surveyors for cost planning at an early stage of the design when there is little detail. Later on, at sketch design stage, unit rates based on similar bill of quantity items from comparable projects are used in conjunction with approximate quantities to produce detailed cost plans.

Cost Planning as a Self-fulfilling Prophecy

Clearly, much apparently firm information is produced from data which may reasonably be regarded as suspect. On the other hand, existing practice seems to keep many professionals happy. The accuracy of estimating both by contractors and professional cost advisors has been the subject of much research in itself. This work has been well reviewed by Ashworth and Skitmore (1982). The most comprehensive of the recent studies shows that the mean deviation of cost planning predictions of the lowest tender is just under 10% (University of Reading 1981). Although far from perfect this figure seems quite good in the light of the quality of the data on which such predictions are based. However it is suggested here that the conventional view of the process of cost planning confuses the issues and a plausible reason is proposed to explain why the accuracy given above is not worse.

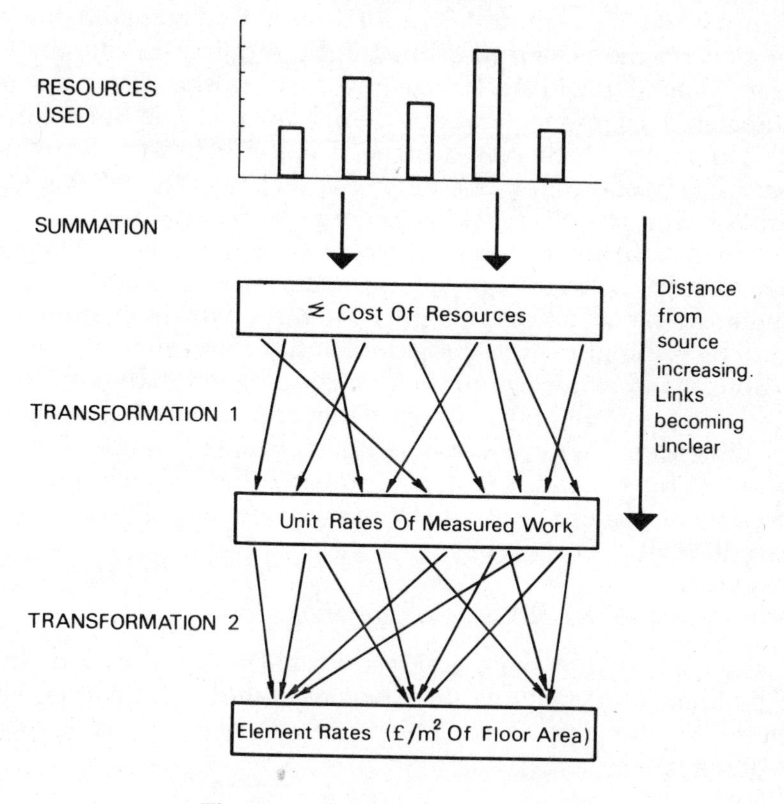

Fig. 11.3 Data transformations

The traditional view of cost planning is that it moves through states of increasing accuracy beginning with fairly low accuracy at the inception stage and then as sketch designs are produced elemental cost planning takes place and eventually cost plans from approximate and then firm quantities. Hence, so the story goes, at the latter stages with much more detailed information, the accuracy of the cost planning (or cost estimating) by the quantity surveyor is much greater that at the beginning. This view assumes that the end point, the accepted tender, is the target identified from the beginning, on which one has eventually 'homed in'. An alternative view which seems to fit the patterns of the process, is that cost planning is a self-fulfilling prophecy.

'Cost planning as a self-fulfilling prophecy' implies that tender figure is fixed from the first time that a budget is given. As the design proceeds more information builds up about the project.

If the project is perceived to be deviating from its pre-planned financial course the control function is exercised and adjustments are made to the design to bring it back into line. Eventually the accepted tender will be sufficiently close to the cost plan and estimates. This prompts the question, how close will the building be to the one which was designed? If the object of the whole exercise is good design, one must guard against the tail wagging the dog. The procedures as they are now, make it very difficult for the cost planner to be shown to be wrong. It is relatively easy to keep the figures in line if the design is altered to suit. A rather extreme instance is given in a report on the Holy Innocents Church, Orpington, the project quantity surveyor said of the accepted competition winning design 'we therefore hypothesised that a building of only half the floor area would be possible'. The dimensions of the Church design were appropriately reduced by the architects and 'the final negotiated figure (including pc's based on quotations) was still marginally higher than the budget' (Farrell 1982).

Data Incompatible with Cost Generation

Cost is generated by the use of resources. The cost data contained in building cost analyses does not represent cost in this way. There are practical reasons for this, however, it is a major inconsistency.

Variability of Data

The coefficients of the variation of bills of quantity rates for similarly described items of work are much greater than figures for whole identical buildings. Beeston (1975) has suggested that the coefficient of variation for identical buildings in the same location would be about 8.5 per cent. He gave typical figures for the trades within bills of quantities as follows: Painter 22%, concretor 15%, steel worker 19%, carpenter 31%, excavator 45%. This would support the idea that bills of quantity rates are merely a notional breakdown of the overall price and are tempered with so many tactical considerations as to render them very unreliable for any form of cost planning.

Consider now the least squares model. This was a global model for a range of building types using global data. A sample of 38 buildings was used to formulate the model. Any assessment of the data here is in effect an assessment of the sample. The sample was statistically small and appeared to be a random sample rather than one designed to extract the maximum of useful information. For example in this case if one wanted a model to fit the case of 'all buildings' the sample would need to be selected so that each member had a weighting in the sample which approximates to its weighting in the population from which the sample is drawn.

In the probabilistic model the data used was that obtained by contactors' estimators. Of all participants in the industry these were the most likely to have access to the best cost data. Each firm needs to keep a detailed record of how and where its costs are incurred merely in order to survive. The potential problems of the recorded data include those referred to earlier. Namely, distortions creeping in at the data collection stage on site particularly with respect to the time taken to perform tasks and material usage and wastage.

Data/Model Interface

The issues of importance at this interface are frequently dependent upon the stage of development of the design at the point at which the model is used. In the context of the models described above the first two are used pre-design and the third probabilistic method is used at the tender stage. In the context of the discussion of data above, it will be seen that all three models perform quite well at the data/model interface. For example the first two models would be commonly used before there are any drawings. At this stage of the design detailed economic data on the current cost of resources such as labour, plant, etc. are of little relevance. What is needed is data of a coarser nature for example the current construction cost per square metre of lettable area. It may also be seen that the Monte Carlo evaluation is well suited for use by contractor's estimators. There is access to detailed cost estimates and a high degree of awareness of the variability of these estimates based on recent company experience.

Finally, in a consideration of the relationship between the model and its data there are at least two further questions. First, is there an available set of data which is at a level of detail appropriate to that of the model? The answer to this will involve a thorough examination to establish exactly what level of detail will enable the maximum benefit to be derived from the model. Secondly, if the model and the data are not well matched it is necessary to establish whether the data is at a greater level of detail than the model. If this is the case then efforts should be directed towards refining the technique of modelling in order to maximise the gain from the existing data. If, on the other hand, the data is at a coarser level of detail than the model, then the relatively larger problems of improving the recorded data need to be addressed.

Model Technique

The technique of the floor area model outlined earlier is difficult to categorise. However, it is suggested that its limitations arise from three features, namely, the comparative nature of the model, linearity and subjectivity.

The floor area model requires that the element unit rate for a previous project be 'adjusted' for time, quality and quantity in order to act as the rate for the project in hand. Any two people using a cost model purely mechanistically should get the same answer if the model is consistent. This is unlikely to be the case with the superficial area model as each of the three adjustments has limitations. The adjustment for time (updating the previous project to today's prices) is carried out by the application of an index. An experiment to evaluate the inaccuracy of indices over time has been described by Raftery (1984b).

The assumptions of linearity in the model relate mainly to the adjustment for quantity. Consider a user preparing a cost plan for an office building in London providing say 8 000 square metres of accommodation. The procedure is that a building analysis is selected which is considered comparable. This is the first limitation, choosing just one building is of course unsound statistically. After updating the analysis by means of tender price index and before making any adjustments for the quality of the specification the element totals are adjusted for the differences in floor area. This straight line adjustment is usually followed by some kind of adjustment to take into account differences in wall to floor ratio, glazing area and so on. The essential problem is that while the straight line adjustment is explicit in the model, the further adjustments are not. There are no clear guidelines for making such adjustments. It is unfortunate that our scales of measurement are not such that we could externalise clearly the 'professional judgement' needed to use the floor area model.

The output of the model is entirely dependent on the building which was chosen as the initial starting point. This may have been a subjective choice. Even if two experienced professionals made precisely the same adjustments for time quality and quantity the answers would be different if they had begun with different buildings.

Surprisingly, these are not complete indictments of the technique. The fact remains that elemental cost planning based on a floor area model is used by professionals, throughout the United Kingdom and in many parts of the world, who seem reasonably happy with the results. The main causes of discomfort may be that significant calculations in the model are not made explicit and are shrouded in professional mystique.

The strengths and weaknesses of the regression technique are well known and will not be discussed in detail here (Gunst and Mason 1980). The Kouskoulos and Koehn model described above produced surprisingly good results despite the limitations of the sample. The coefficient of multiple correlation, which among other things measures the closeness of fit of the regression plane to the points, was = 0.998, which indicates a very high degree of correlation. In an interesting aside the authors pointed out that the most subjective variables, 'quality' and 'technology', were also the ones with the highest correlation coefficients. This shows mathematically that the shrouded area of professional experience discussed above with respect to the first model, is indeed an extremely important factor in cost and price forecasting. This particular regression model was surprisingly robust considering that it was designed to be used pre-design, i.e. before introduction of drawings.

In the construction of a model for Monte Carlo evaluation there are two special requirements. The first is that the model be constructed in such a way that the main variables are independent. In the context of a probabilistic estimate of building cost this means that the building needs to be broken down into reasonably independent sub-systems. The probabilistic approach demonstrated above may have appeared to be crude but it was designed to take into account this requirement. A vast amount of information relating to the project cost is processed by the estimating team during the tendering stage. It is relatively easy to break down the estimate into dozens of items and sub-items. This process of breaking the process down into smaller and smaller components is known as desegregation. However, as desegregation increases so does the likelihood of correlation among the variables, so in using Monte Carlo evaluation it is important to resist the temptation to increase the level of desegregation.

The second, and less important, requirement is knowledge of the probability distribution of each variable. Given that there are no large statistical samples of the distributions of building costs from which one could establish the 'true' distribution for each variable. In the case described above, the estimators' subjective perceptions of probability distributions were mapped during a series of workshops with the contractors' estimators. The distributions were regarded as project-specific, this meant that if it was required to carry out a risk analysis of a particular cost estimate prior to making a competitive tender, the estimating team would spend one to two hours defining their perceptions of the probability distributions for the sub-systems for that particular project. These distributions were not used for other projects. The distributions for each project were built up from first principles although the experience gained in initial projects meant that the time taken to define the subjective perceptions of probability reduced on later projects.

Finally, it has been shown conclusively that correlation effects have more serious consequences than the choice of probability distribution (Pouliquen 1970).

Interpretation of Output

From the point of view of the decision maker, given a knowledge of the model technique, the data/model interface and the data, there are at least two key issues remaining. These are the interpretation of the model output with respect to firstly, the age of the model, and secondly, the uncertainty in the model data and output. As models age there are three ways in which they may be updated. They may be entirely reformulated with new relationships and new data. The model may be left as it stands but the data files updated. The model and data may both be left as they are and the output may be adjusted by the application of an index.

Both the input data and the output from the floor area model are updated by means of cost and price indices. The output is usually a deterministic figure, 'the estimated tender price for this building, one year hence, will be £x'. From the point of view of a decision-maker this is a rather inflexible figure. There is no statistical description of the result. No estimate of the possible range of building costs and no estimate of the chances of that particular cost being exceeded. For these reasons the actual output achieved is frequently a conservative one. It was pointed out earlier that, if as the design progresses it is perceived that the target is being missed, the target is then conveniently moved (by redesigning the building).

Least squares models of the type described earlier give a similar deterministic output with one important difference, this is that the nature of the model is such that a statistical description of the model's strengths and weaknesses is available to the decisionmaker. He or she will not only have access to the coefficient of multiple correlation but can also rank the input variables in order of their correlation with the output and test the sensitivity of the model to changes in, or deletions of, particular variables. Given the constraints of the least squares method, mainly extrapolation and causation issues, the output accompanied by appropriate statistics is an extremely important decision-making tool.

The output from Monte Carlo evaluation is usually given in the form of frequency histograms and cumulative probability diagrams. Thus, it is usually possible to specify a given degree of risk and read off an appropriate output. For example, one could say 'I am going to use an estimate, which given all these assumptions, has only a 15% chance of being exceeded'. It was pointed out earlier that first, second and third generation models may be seen as products of their time and the thinking of their time. It should be pointed out that probabilistic estimates or risk analyses do not diminish the risks, remove the risks or control the risks. They merely attempt to evaluate them in some reasonably consistent numeric fashion. In essence, they attempt to attach numbers to subjective perceptions of risks in estimates of future events.

Clearly, correct interpretations of output by decision makers can only be made if the decision makers are familiar with the strengths and weaknesses of the model technique employed. One of the disadvantages of increasingly sophisticated computer software is that the decision maker is becoming more and more removed from the model technique. It is to be hoped that by applying the four criteria used here, a decision maker will be able to gain a sufficiently systematic overall view of a cost or price model in order to be able to interpret the output in some meaningful way.

Conclusions

This chapter has attempted to demonstrate a commonsense way of quickly evaluating the strengths and weaknesses of models of building cost or price using four simple systems criteria. The choice of three models used for the evaluation suffers from a fundamental limitation. They are not mutually exclusive or representative of a logical categorisation of attributes. Philosophically, perhaps one should have chosen a procedural model, a statistical model, an optimisation model and a stochastic model, (classified by technique) or a cost model and a price model (classified by output). Although logically sound, this would perhaps have done a disservice to the more pressing concerns of professionals in the field for whom the three models taken above are topical and representative of the range of techniques in current use. A comprehensive review of cost models is given in Marshall (1990a, 1990b).

What then is the state of building cost and price modelling? Although in this short essay it has only been possible to give the briefest overview, it would appear that in this particular field, the problems of model technique and data/model interface are not major. It is commonly perceived that the quality of cost data available to professionals in the industry is not good. This may well be, but it should also be remembered that the industry is characterised by a high degree of uncertainty. The advent of onsite computers means that the organisational and managerial changes necessary to improve the quality of data collection on site are less of a barrier than they used to be. However, given the nature of competitive tendering and the organisation of the industry in general, design teams, cost consultants and quantity surveyors will not be able to improve their access to data in the foreseeable future. Given that this is the case (the second assumption is arguable) it could be said that the models are 'no better than they ought to be'. However, the problem which needs to be addressed both by researchers and practitioners in the field, is how to make better use of the data and models which already exist. The problem here, it is suggested, is one of education. Professionals and academics need to be trained to question consistently and systematically the assumptions and to require, as a matter of course, statistical descriptions of all estimates.

Interestingly, the most commonly used pre-design model in the UK is still the first generation floor area model. Although there is emerging evidence that regression methods can outperform such manual approaches (Wills and Raftery 1990). The contribution of second and third generation models is important however, in that they have forced academics and professionals alike to seek ways to make better use of the existing cost information. Professionals and clients are most familiar with the first generation floor area model. Therefore one logical course of action would be to incorporate the experience from second and third generation models into an improved version of this first generation model. This has already begun with the computerised versions of elemental cost planning and could be continued by an increasing familiarity with and use of statistical descriptors, probabilistic estimates and expert systems.

Given a set of drawings, a cost consultant or quantity surveyor is able to predict tender price with an acceptable degree of uncertainty. The most difficult areas of modelling are, from the point of view of the design team, the forecasting of building price before drawings are produced, and from the point of view of building contractors, the evaluation, in today's turbulent building environment, of financial and construction risks on projects.

Bibliography

Abel C. (1980) 'Meaning and rationality in design.' In *Meaning and Behaviour in the Built Environment*, (Ed. by G. Broadbent, R. Blunt and T. Lorens) Wiley, Chichester.

Alexander C. (1963) The determination of components for an Indian village. In *Conference on Design Methods*, (Ed. by J. C. Jones and D. G. Thornley) Pergamon, Oxford, pp. 83–114.

Alexander C. (1964) *Notes on the Synthesis of Form*. McGraw-Hill, New York.

Alexander C. (1965) *The co-ordination of the urban rule system*. Working Paper No. 1, Centre for Planning and Development Research, University of California.

Alexander C. (1966) 'A city is not a tree'. *Design*, 206, pp. 44–55.

Alexander C. (Dec 1971) *'What was design methodology daddy?'*, An interview with Christopher Alexander, Architectural Design.

Alexander C. *et al.* (1975) *The Oregon Experiment*. Oxford University Press, New York.

Alexander C. *et al.* (1977) *Pattern Language: Towns, Buildings and Construction*. Oxford University Press, New York.

Alexander C. *et al.* (1979) *Timeless Way of Building*. Oxford University Press, New York.

ASCE, (1917) Committee on Valuation of Public Utilities, The life experience of 14 railway stations. *Transactions of the ASCE*, **XXXI**, December 1917, p. 1557.

Ashworth A. and Skitmore M. (1982) *Accuracy in Estimating*. Occasional Paper No. 27, Chartered Institute of Building.

Au T. and Hendrickson C. (1986) 'Profit Measures for construction Projects.' *Journal of the ASCE, Construction Division*, **112** (2), June, pp. 273–286.

BPF (1983) *Manual of the B P F System for Building Design and Construction*. The British Property Federation, London.

Balchin P. N. (1981) *Housing Policy and Housing Needs*. Macmillan, London.

Balchin P. N. (1990) *Regional Economics: The North–South Divide*. Paul Chapman, London.

Balchin P. N., Kieve G. and Bull G. (1988) *Urban Land Economics and Public Policy*. 4th edn, Macmillan, London.

Ball M. (1988) *Rebuilding Construction: Economic Change in the British Building Industry*. Routledge, London.

Barber W. J. (1967) *A History of Economic Thought*. Penguin, New York.

Barron M. and Targett D. (1985) *The Managers' Guide to Business Forecasting*. Basil Blackwell, Oxford.

Bathurst P. and Butler D. (1973) *Building Cost Control Techniques and Economics*, Heinnemann, London.

Baumol W. J. (1959) *Business Behaviour, Value and Growth*. Macmillan, New York.

Beeston D. (1975) One statisticians view of estimating. *The Building Economist*, December, pp. 139–145.

Benjamin N. B. H. and Meador R. C. (1979) Comparison of Friedman and Gates competitive bidding models. *Journal of the Construction Division, ASCE*, March, pp. 25–40.

Berle A. A. and Means G. C. (1932) *The Modern Corporation and Private Property*. Commerce Clearing House Inc., New York.

Bernoulli D. (1954) Exposition of a new theory on the measurement of risk. *Econometrica*, **22**: 23–6. (Translated from the original Latin by L. Summer, first published in 1738.)

Bishop D. (1975) Productivity in the construction industry. In *Aspects of the Economics of Construction*, (Ed. D. A. Turin) pp. 59–96, George Goodwin, London.

Bon R. (1986a) Choices, Values and Time: The psychology of cost-benefit assessments. *Building Research and Practice*, **14**(4), 223–5.

Bon R. (1986b) Timing of space: some thoughts on building economics. *Habitat International*, **10**, (4), 101–7.

Bon R. (1989) *Building as an Economic Process: An Introduction to Building Economics*. Prentice Hall, Englewood Cliffs, New Jersey.

Boskin M. (1987) *Reagan and the Economy*. International Centre for Economic Growth, San Francisco.

Bowen P. (1982a) An alternative estimating approach. *Chartered Quantity Surveyor*, February, pp. 191–4.

Bowen P. (1982b) Problems in econometric cost modelling. *The Quantity Surveyor*, May, pp. 83–5.

Bowen P. (1984) 'Applied Econometric Cost Modelling'. CIB Working Commission on W55, *Proceedings Third International Symposium on Building Economics*, Ottowa, **3**, pp. 144–157.

Bowen P. A. and Edwards D. J. (1985) Cost modelling and price forecasting: practice and theory in perspective. *Construction Management and Economics*, **3**, No. 3, pp. 199–215.

Brandon P. A. and Newton S. (1985) Improving the forecast. *Chartered Quantity Surveyor*, May, pp. 14–26.

Briscoe G. (1988) *The Economics of the Construction Industry*. Mitchell, London.

Bucciarelli Louis L. (1988) An ethnographic perspective on engineering design. *Design Studies*, **9** (3), pp. 159–168.

Buchanan J. S. (1973) *Cost Models for Estimating*, Royal Institute of Chartered Surveyors, London.

Building Market Report (1989) *High-tech goes ex-growth*. *Building Market Report*, September, 8.

Canter D. (1977) *The Psychology of Place*. Architectural Press, London.

Canter D. (1984) Beyond building utilisation. In *Designing for Building Utilisation*, (Ed. by J. A. Powell, I. Cooper and S. Lera), E. & F. N. Spon, London. pp. 41–7.

Chapman J. J. and Jones M. (1981) *Models of Man*, British Psychological Society, London.

Chartered Institute of Building (1983) *Code of Estimating Practice*. 5th edn. Chartered Institute of Building.

Chau K. W. and Walker A. (1988) The measurement of total factor productivity of the Hong Kong construction industry. *Construction Management and Economics*, 6, pp. 209–224.

Clapp M. A. (1963) Cost comparisons in housing maintenance. *Local Government Finance*, **67**, October.

Clapp M. A. and Cullen B. (1968) The maintenance and running costs of school buildings. *Chartered Surveyor*, May, pp. 552–560.

Clarke R. and McGuiness T. (1985) *The Economics of the Firm*. Basil Blackwell, London.

Cooper D. F. and Chapman C. B. (1987) *Risk Analysis for large Projects: Modes, Methods and Cases*. John Wiley and Sons, London.

Cormican D. (1985) *Construction Management: Planning and Finance*. Construction Press, London.

Cowan P. C. (1965) Depreciation, Ageing and Obsolescence. *Architects Journal*, **141**, (24), June, pp. 1395–1401.

Cowan P. and Sears J. (1966) *Growth, Change, Adaptability and Location*, (mimeo.), Joint Unit for Planning Research, London.

Cowie H., Harlow C. and Emerson R. (1984) *Rebuilding the Infrastructure*. Policy Studies Institute, London.

Cross N., Naughton J. and Walker D. (Oct 1981) Design Method and Scientific Method. *Design Studies*, **2** (4), pp. 195–202.

Cullen B. and Jeffrey I. (1967) *Running Costs of Hospital Buildings*. Building Research Station, Current Paper, Design Series, No. 65, HMSO, London.

Cyert R. M. and March J. G. (1963) *A Behavioural Theory of the Firm*. Prentice Hall, Englewood Cliffs, New Jersey.

Dasgupta A. K. *Epochs of Economic Theory*. Blackwell, London.

Davis, Belfield and Everest (Eds) (1988) *Spon's International Construction Costs Handbook*, E. and F. N. Spon, London.

Dell'Isola A. J. and Kirk S. J. (1981) *Life Cycle Costing for Design Professionals*. McGraw-Hill, New York.

Dixie J. M. (1974) Bidding Models – the final resolution of a controversy. *Journal of the Construction Division, ASCE*, September, pp. 265–271.

Dobb M. (1973) *Theories of Value and Distribution since Adam Smith*. Cambridge University Press, London.

DOE (1971a) *Costs in Use: A Study of 24 Crown Office Buildings*. Department of Quantity Surveying Development, HMSO, London.

DOE (1971b) *Costs in Use – A Guide to Data and Techniques*. HMSO, London.

DOE (1977) 'Occupancy costs of offices'. *DOE Construction*, No. 23, September, pp. 5–8.

Downie J. (1958) *The Competitive Process*. Duckworth, London.

Drabkin D. H. (1977) *Land Policy and Urban Growth*. Pergamon Press, London.

Eastman C. (1970) On the analysis of intuitive design processes. In *Emerging Methods in Environmental Design and Planning*, (Ed. G. T. Moore), MIT Press, Boston.

Eccles R. G. (1981) The quasifirm in the construction industry. *Journal of Economic Behaviour and Organisation*, **2**, 335–357.

The Economist Newspaper (1988) *Britains Supply Side Miracle*. December 3, pp. 103–4.

The Economist Newspaper (1989) *Growth can be green*. August 26, p. 14.

Elkins P. (Ed) (1986) *The Living Economy*. Routledge & Kegan Paul, London.

Elkins P. (1988) Mapping Out a Living Economy. New Economics **6**, Summer, 3–5.

Engelbrecht-Wiggans R. (1980) Auctions and Bidding Models: A Survey. *Management Science*, **26**, pp. 119–142.

Eppink D. (1975) *Flexibility: a necessary complement to planning*.

European Institute for Advanced Studies in Management, Working paper, 1975–24.

Farrell (1982) Architecture for innocents. *Chartered Quantity Surveyor*, January, p. 152.

Fellows R. (1988) *Escalation Management in Construction*. Unpublished PhD Dissertation, University of Reading.

Ferry D. J. and Brandon P. S. (1991) *Cost Planning of Buildings*. Sixth edn. BSP Professional Books, Oxford.

Fitzroy F. and Mueller D. C. (1984) Cooperation and Conflict in Contractual Organisations. *Quarterly Review of Economics and Business*, **24**, pp. 24–50.

Flanagan R. and Norman G. (1983) *Life Cycle Costing for Construction*. RICS, London.

Flanagan R. and Norman G. (1989) 'Pricing Policy'. In *The Management of Construction Firms* (Ed. by Hillebrandt and Cannch pp. 129–153.

Flanagan R., Norman G., Meadows J. and Robinson G. (1989) *Life Cycle Costing*. BSP Professional Books, Oxford.

Foster N. In Suckle A. *ibid* p. 138.

Frank J. (1981) *Building Procurement Systems*, Chartered Institute of Building, Ascot.

Friedman L. (1956) A Competitive Bidding Strategy. *Operations Research*, **14**, pp. 104–112.

Friedman M. (1968) The role of monetary policy. *American Economic Review*, **1**, pp. 1–17.

Friedman M. (1982) *The Federal Reserve & Monetary Instability*. Wall Street Journal, February 1st.

Gates M. (1967) Bidding Strategies and Probabilities'. *Journal of the Construction Division, ASCE*, March, **93**, pp. 75–103.

Georgescu-Roegen N. (1971) *The Entropy Law and The Economic Process*. Harvard University Press, Cambridge, Mass.

Gill R. (1980) Approaches to Design. *Design Studies*, **1**(3), pp. 141–5.

Gordon A. (1974) *The Economics of the 3L's Concept*. Research and Development Paper 8, Royal Institution of Chartered Surveyors, London.

Gould P. R. (1970) *The development of a cost model for the heating ventilating and air-conditioning installation of a building*. MSc project report, Loughborough University.

Groome C. (1978) *Construction for Industrial Recovery*. Building and Civil Engineering EDC, HMSO, London.

Gunst R. F. and Mason R. L. (1980) *Regression Analysis and Its Application*. Marcel Dekker, New York.

Gwynn J. (1968) Package deal threat. *Consulting Engineer*, (**32**), July, pp. 59–64.

Hall R. L. and Hitch C. J. (1939) Price theory and business behaviour. *Oxford Economic Papers*, **2**, pp. 12–45.

Handy C. (1988) Re-inventing the place of work. *New Economics* 4, Winter, pp. 2–5.

Haworth J. G., (1973) Externalities and the city. In *The Modern City: Readings in Urban Economics*, (Ed. by D. W. Rasmussen and C. T. Haworth) pp. 28–30, Harper and Row, London.

Hayek F. A. (1960) *The Constitution of Liberty*. Routledge and Kegan Paul, London.

Hayes R. W. *et al*. (1983) *Risk Management in Engineering Construction*. SERC Research Report, Thomas Telford, London.

Henderson D. (1985) *Innocence and Design: The Influence of Economic Ideas on Policy*. Blackwell, London.

Hertz D. B. (1979) Risk analysis in capital investment. HBR Classic, *Harvard Business Review*, Sept/Oct, pp. 169–182.

Hertz D. B. and Thomas H. (1983) *Risk Analysis and its Applications*. Wiley, London.

Hewison R. (1987) *The Heritage Industry: Britain in a climate of decline*. Methuen, London.

Hicks J. R. (1948) *Value and Capital: An inquiry into some fundamental principles of Economic Theory*. Oxford University Press, London.

Hicks J. R. (1973) Capital and Time: a Neo-Austrian Theory. Clarendon Press, Oxford.

Higgin G. and Jessop N. (1965) *Communications in the Building Industry*. The Tavistock Institute, London.

Hillebrandt P. M. (1984) *Analysis of the British Construction Industry*. Macmillan, London.

Hillebrandt P. M. (1985) *Economic Theory and the Construction Industry*, 2nd Edition, MacMillan, London.

Hillebrandt P. M. and Cannon J. (Eds) (1989) *The Management of Construction Firms*. Macmillan, London.

Hillebrandt P. M. and Cannon J. (1989) *The Modern Construction Firm*. Macmillan, London.

Holmstrom B. (1979) Moral hazard and observability. *Bell Journal of Economics*, **10**, pp. 74–91.

Humphreys C. (1976) *Zen Buddhism*, Allen & Unwin, London.

Institution of Civil Engineers (ICE) (1976) *An Introduction to Engineering Economics*. ICE, London.

Institution of Civil Engineers (1988) *Overseas Projects: Crucial Problems*. Thomes Telford, London.

Jefferies-Matthews E. D. (1970) If we must have package deals. *Construction News*, March 19, p. 10.

Jenkins G. M. (1978) *Practical Experiences in Business Forecasting*. S. M. Jenkins & Partners.

Jensen M. (1983) Organisation Theory and Methodology. *Accounting Review*, **58**, pp. 319–339.

Jones J. C. (1970) *Design Method, Seeds of Human Futures*. Wiley, London.

Jones J. C. and Thornley D. G. (1963) *Conference on Design Methods*. Pergamon, Oxford.

Kahneman D. and Tversky A. (1984) Choices, values and frame. *American Psychologist*, **39**(4), April, pp. 71–79.

Kamarck A. (1983) *Economics and the Real World*. Basil Blackwell, London.

Keynes J. M. (1936) *The General Theory of Employment, Interest and Money*. MacMillan, London.

King M. and Mercer A. (1988) Recurrent Competitive Bidding. *European Journal of Operational Research*. **33**, pp. 2–16.

Kirwan R. and Martin D. B. (1972) *The Economics of Urban Residential Renewal and Improvement*. Working Paper No. 77, Centre for Environmental Studies, London.

Kouskoulos V. and Koehn E. (1974) Predesign Cost Estimation Function for Building, *ASCE, Journal of the Construction Division*, December, pp. 589–604.

Koffka K. (1935) *Principles of Gestalt Psychology*. Harcourt Brace, New York.

Kostoff S. (Ed.) (1977) *The Architect: Chapters in the History of the Profession*. Oxford University Press, New York.

Krauss R. I. *et al.* (1970) *Improving Design Decisions*. (Unpublished) Ministry of Public Building and Works, London.

Landreth H. (1976) *History of Economic Theory*. Houghton Mifflin, Boston.

Langford D. A. (1982) *Direct Labour Organisations in the Construction Industry*. Gower, Aldershot.

Lansley P. R. (1987) Corporate strategy and survival in the UK construction industry. *Construction Management and Economics*, **5**, pp. 141–155.

Lawson B. (1980) *How Designers Think*. Architectural Press, London.

Lean W. and Goodall B. (1966) *Aspects of Land Economics*. Estates Gazette, London.

Leibenstein H. (1966) Allocative Efficiency vs X-Efficiency. *American Economic Review*, **56**, June.

Lera S. (1983) Synopses of some recent published studies of the design process and designer behaviour. *Design Studies*, **4** (2), pp. 133–140.

Lewis J. P. (1979) *Urban Land Economics: A set approach*. Edward Arnold, London.

Lewis T. M. (1987) Is Productivity a problem?, In *Managing Construction World Wide*, (Ed. by P. Harlow and P. Langley) pp. 778–787 E. and F. N. Spon, London.

Lowe J. R. (1970) The consultant's case against package deals. *Construction News*, January 8, p. 18.

Lowe J. G. (1987) The measurement of productivity in the construction industry. *Construction Management and Economics*, **5**, pp. 101–113.

Lynch K. (1972) *What Time is This Place*. MIT Press, Boston.

McCaffer R. (1975) Some examples of the use of regression analysis as an estimating tool. *The Quantity Surveyor*, December, pp. 57–75.

McGuiness T. (1987) Markets and managerial hierarchies. In *The Economics of the Firm*, (Ed by R. Clarke & T. McGuiness) pp. 42–61. Blackwell, London.

Mackinder M. and Marvin H. (1982) *Design Decision Making in Architectural Practice*. York IAAS, Research Paper 19, Institute of Advanced Architectural Studies, University of York, p. 114.

Malcolmson J. (1984) Efficient labour organisation: incentives, power and the transaction cost approach. In *Firms Organisation and Labour: Approaches to the Economics of Work Organisation*, (Ed by F. H. Stephen) Macmillan, London.

Male S., Stocks R. and Torrence V. (Eds) (1989) *Competitive Advantage in Construction*, Butterworths London.

Malthus T. R. (1959) *Population – the First Essay*. Michigan University Press, Michigan.

Manser W. A. P. (1985) *The British Economic Base 1985*. Federation of Civil Engineering Contractors, London.

March L. J. (1976) The logic of design and the question of value. *The Architecture of Form*, (Ed. by L. J. March), Cambridge University Press, London.

Markowitz H.M. (1959) *Portfolio Selection: Efficient Diversification of Investments*. Wiley, New York.

Markus T. A. *et al*. (1972) *Building Performance*. Applied Science, London.

Marris R. (1964) *The Economic Theory of Managerial Capitalism*. Macmillan, London.

Marshall A. (1961) *Principles of Economics*. 9th (variorum) edn. (Ed by C. W. Guillebaud), Macmillan, London. (First published 1890.)

Marshall H. E. (1989) Review of Economic Methods and Risk Analysis Techniques for evaluating building investment (Part 1). *Building Research and Practice*, **6**, December, pp. 242–9.

Marshall H. E. and Ruegg R. T. (1981) *Recommended Practice for Measuring Benefit/Cost and Savings-to-investment Ratios for Buildings and Building Systems*. National Bureau of Standards, Interagency Report 81–2397 November. National Bureau of Standards, Washington, D.C.

Marx K. (1976) *The Process of Production of Capital – A Critique of Political Economy*. **1**, Penguin, Hamondsworth.

Marx K. (1956) *The Process of Circulation of Capital – A Critique of Political Economy*. **2**, Lawrence & Wishalt, London.

Marx K. (1959) *The Process of Capitalist Production as a Whole – A Critique*. **3**, Lawrence & Wishalt, London.

Matchett E. (1968) Control of thought in creative work. *The Chartered Mechanical Engineer*, **15**, pp. 163–166.

Maver T. W. (1970) A theory of architectural design in which the role of the computer is identified. *Building Science*, **4** (4), pp. 199–208.

Maver T. W. (1977) Building appraisal. In *Computer Applications in Architecture*, (Ed. by J. S. Gero), Applied Science, London. pp. 63–94.

Max-Neef M. (1986) Human-scale economics: the challenges ahead. In *The Living Economy*, (Ed. by P. Elkins), pp. 45–54. Routledge, London.

Meade J. (1975) *The Intelligent Radicals Guide to Economic Policy*, George Allen and Unwin, London.

Medhurst D. F. and J. P. Lewis (1969) *Urban Decay: An analysis and policy*, Macmillan, London.

Menger C. (1981) *Principles of Economics*, New York University Press, New York and London, (first published in 1871; first English translation published in 1950).

Mill J. S. (1970) *Principles of Political Economy with Some Other Applications to Social Philosophy*, Penguin, Harmondsworth.

Miller A. (1987) Blueprint for a school of economic thought. *New Economics*, No. **2** June pp. 4–5.

Minford A. P. L. and Peel D. (1983) *Rational Expectations and the New Macroeconomics*, Martin Robertson, London.

Mishan E. J. (1990) Economic and political obstacles to environmental sanity. *National Westminster Bank Quarterly Review*,

May, pp. 25–42.
Morris D. (1985) *The Economic System in the UK*, Oxford University Press, London.
Morrison N. (1984) The accuracy of quantity surveyors cost estimating. *Construction Management and Economics*, **2**, pp. 57–75.
Mulligan J. (1989) *Managerial Economics*, Allyn and Bacon, Needham Height, Mass.
National Economic Development Office (1976) *Cyclical Fluctuations in the United Kingdom Economy*, NEDO Books, London.
Newland P., Powell J. A. & Creed C. (1987) Understanding architectural designers' selective information handling. *Design Studies*, **7**, No. 1, January, pp. 2–16.
Niskanen W. (1988) *Reaganomics*. OUP, New York.
Nutt B. *et al*. (1976) *Obsolescence In Housing*, Saxon House, London.
Nyman S. & Silbertson A. (1978) The ownership and control of industry. In *Readings in Applied Microeconomics*, 2nd Edn. (Ed. by L. Wagner), pp. 214–240 Open University Press, London.
The Other Economic Summit (1984) *Report and Summary of the Other Economic Summit*, June, 6–10.
The Other Economic Summit (1986) *Report and Summary of the Other Economic Summit*, TOES, Tokyo.
Park W. (1979) *Bidding For Profit*, Wiley, New York.
Pearce D., Markandaya A. & Barbier E.B. (1989) *Blueprint for a Green Economy*, Earthscan Publications Ltd, London.
Pepinster C. (1989) Collaring the artful dodgers. *Building*, 1st Sept. pp. 14–16.
Pilcher R. (1984) *Project Cost Control in Construction*, Collins, London.
Popper K. (1964) *Conjectures and Refutations*, Gollancz, London.
Poincare H. (1929) *The Foundations of Science*, Science House Inc., New York.
Posner M. I. (1973) *Cognition, an Introduction*, Scott Foresman Basic Psychological Concept Series, London.
Pouliquen L. Y. (1970) *Risk Analysis in Project Appraisal*, World Bank Staff Occasional Paper No. 11, John Hopkins Press, Baltimore.
Powell C. G. (1982) *An Economic History of the British Building Industry 1815–1979*, Methuen, London.
Raftery J. (1984a) 'Models in building economics: a conceptual framework for the assessment of performance', CIB Working Commission W55, *Proceedings Third International Symposium on Building Economics*, Ottawa, **3** pp. 103–111.

Raftery J. (1984b) *An investigation into the suitability of cost models for use in building design*. PhD. Thesis, Liverpool Polytechnic, CNAA.

Raftery J. (1985a) *Strategies for Improving the Performance of Finnish Firms Bidding in the International Construction Market*, (mimeo), Technical Research Centre of Finland, Laboratory of Building Economics, Helsinki.

Raftery J. (1985b) Riskianalyysi tarjouslaskennan apuna, (Risk Analysis as a bidding aid), *Rakennus tekniikka*, Finland, **6**, pp. 439–440.

Raftery J. (1985c), *Risk Analysis in Construction Tendering – A Guide for Building Contractors*, Technical Research Centre of Finland, Laboratory of Building Economics p. 44.

Raftery J. (1988) Dynamic rehabilitation. *RIBA Journal*, August, pp. 61–77.

Ramsay W. (1989) 'Business objectives and strategy.' In *The Modern Construction Firm*, (Ed. by P. M. Hillebrandt and J. Cannon), pp. 9–29. Macmillan, London.

Reutlinger S. (1970) *Techniques for Project Appraisal under Uncertainty*, World Bank Staff, Occasional Papers, No. 10, John Hopkins Press, Baltimore.

RIBA (1965) *Architectural Practice and Management Handbook*, RIBA Publications, London.

Rogers R. (1980) In *By Their Own Design*, A. Suckle, Granada, p. 112.

Rosenshine M. (1972) Bidding models: resolution of a controversy. *Journal of the Construction Division, ASCE*, March, pp. 143–148.

Ruegg R. T. (1984) *Economic Evaluation of Building Design, Construction, Operation and Maintenance*, U.S. Department of Commerce, Technical Note No. 1195, National Bureau of Standards, Gaithersburg, Maryland.

Ruegg R. T., McConnaughey J. S., Sav G. T. & Hockenberry K. A. (1978) *Life-Cycle Costing: A Guide for Selecting Energy Conservation Projects for Public Buildings*, National Bureau of Standards Building Science Series 113, National Bureau of Standards, Washington, D.C.

Ricardo D. (1973) *The Principles of Political Economy and Taxation*, Dent, London.

Ruegg R. T. (1982) *Life-cycle Cost Manual for the Federal Energy Management Program*, National Bureau of Standards Handbook Series 135 (Rev.), National Bureau of Standards, Washington, D.C.

Salway F. (1986) *Depreciation of Commercial Property*, CALUS Research Report, Centre for Applied Land Use Studies, Reading.

Samuelson P. A., Nordhaus W. D. (1985) *Economics*, 12th edn McGraw-Hill, New York.

Schon D. A. (1988) Designing: rules, types and worlds. *Design Studies*, **9**, No. 3, July pp. 181–190.

Schumpeter J. A. (1954) *History of Economic Analysis*, Allen and Unwin, London.

Scitovsky T. (1978) *The Joyless Economy: An inquiry into human satisfaction and consumer dissatisfaction*. OUP, Oxford, London, New York.

Seeley I. H. (1983) *Building Economics*, 3rd edn Macmillan, London.

Selsdon L. & Palmer R. H. (1988) Financial problems – a bankers view. In *Overseas Projects – Critical Problems*, (Institution of Civil Engineers), pp. 45–8. Thomas Telford Ltd., London.

Shackle G. L. S. (1961) *Decision, Order and Time in Human Affairs*, Cambridge University Press, Cambridge.

Silveira J. (1971) *Incubation: The effect of interruption, timing and length on problem processing*, (unpublished) PhD. dissertation, University of Oregon.

Simmonds R. (1980) Limitations in the decision strategies of design students. *Design Studies*, **1** No. 6, pp. 358–364.

Skitmore M. (1988) *Fundamentals Research in Bidding and Estimating*, (mimeo), presented to CIB W55, Building Economics Workshop, May 1988, Haifa.

Skitmore M. (1989) An introduction to bidding strategy. (Ed. by S. P. Male, R. K. Stocks & V. B. Torrance) In *Competitive Advantage in Construction*, Butterworths, London.

Smith A. (1776) *The Wealth of Nations*, (Ed. by Cannan E. (1961), Methuen, London.

Smith S. & Wiede-Nebbeling S. (1986) *The Shadow Economy in Britain and Germany*, The Anglo German Foundation, London.

Stewart M. (1972) *Keynes and After*, 2nd ed., Pelican, London.

Stone P. A. (1979) *Urban Development in Britain: Standards, Costs and Resources, 1964–2004*, Cambridge University Press, London.

Stone P. A. (1980) *Building Design Evaluation: Costs in Use*, Spon, London.

Stone P. A. (1983) *Building Economy*, 3rd edn Pergamon, London.

Strassman W. P. & Wells J. (Eds) (1988) *The Global Construction Industry: Strategies for Entry, Growth and Survival*, Unwin Hyman, London.

Strong N. & Waterson M. (1987) Principals, agents and information. In *The Economics of the Firm*, (Ed. by R. Clarke & T. McGuiness), pp. 18–41. Blackwell, London.

Suckle A. (1980) *By Their Own Design*, Granada, New York.

Sugden J. D. (1978) Direct Labour: how productivity statistics have proved nothing. *Municipal Engineering*, pp. 354–71.

Sugden J. (1980) The nature of construction capacity and entrepreneural response to effective demand in the UK. *The Production of the Built Environment*, **1**, pp. 1–6, Bartlett School of Architecture and Building, University College, London.

Sweezy P. M. (1946) *The Theory of Capitalist Development*, Denis Dobson, London. 2nd edn.

Switzer J. F. Q. (1963) The Life of Buildings In An Expanding Economy. *Chartered Surveyor*, August, pp. 70–7.

Thurow L. C. (1983) *Dangerous Currents: The State of Economics*, Oxford University Press, London.

Thurow L. C. (1987) *The Zero-Sum Solution*, Revised edition, Penguin, London.

Turner D. (1987) The Construction Industry in Britain. *Midland Bank Review*, Autumn, pp. 16–23.

Turunen I. (1984) Modern methods for cost estimation and decision making in preliminary plant design. *Acta Polytechnica Scandinavica*, Chemical Technology and Metallurgy Series, No. 155.

Tversky A. & Kahneman D. (1975) Judgement Under Uncertainty: Heuristics and Biases. *Science*, **185**, 1975, 1124–1131.

University of Reading (1981), *Cost Planning and Computers*, Department of Construction Management, University of Reading.

Vergara A. J. & Boyer L. T. (1977) Portfolio theory: applications in construction. *Proceedings of the American Society of Civil Engineers, Journal of the Construction Division*, March, pp. 23–38.

Vergara A. J. (1977) *Probabilistic estimating and applications of portfolio theory in construction*. PhD. thesis, University of Illinois, Urbana.

Walras L. (1954) *Elements of Pure Economics*, translated by William Jaffe, Allen and Unwin, London, (first published in French in 1870).

Ward R. (1987) Office building systems performance and functional use costs. CIB Working Commission W55 Building Economics, *Proceedings of the Fourth International Symposium on Building Economics*, Copenhagen, vol. **A** pp. 113–124.

Ward B. and Dubos R. (1972) *Only One Earth – The Care and Maintenance of a Small Planet*, Report for UN Habitat, Andre Deutsch, London.

Weber S. F. & Lippiatt B. C. (1983) *Productivity Measurement for the Construction Industry*, National Bureau of Standards, Technical Note 1172, Department of Commerce, Washington D.C.

Wiles R. M. (1976) Cost Model for a Lift Installation. *The Quantity Surveyor*, May.

Williams N. (1981) *Influences on the Profitability of 22 Industrial Sectors*, Bank of England, Discussion Paper No. 22, Bank of England, London.

Williams A. (1985) *Obsolescence and Re-use: A Study of Multi–storey Industrial Buildings*, (mimeo), Report for Hunter and Partners Educational Trust, School of Land and Building Studies, Leicester Polytechnic.

Williamson O. E. (1967) *The Economics of Discretionary Behaviour: Managerial Objectives in a Theory of the Firm*, Markham Publishing Company, Chicago.

Wills D. J. & Raftery J. (1990) *Modelling the Cost of Air Conditioning Systems*, In press.

Wilson A. J. (1982) Experiments in Probabalistic Cost Modelling. In *Building Cost Techniques, New Directions*, (Ed. by P. S. Brandon), pp. 169–180. E. & F. N. Spon, London.

Wittkower R. (1962) *Architectural Principles in the Age of Humanism*, Alec Tiranti Ltd, London.

Wood E. (1972) *Property Investment: A real value approach*, (unpublished) PhD. Thesis, University of Reading.

World Bank (1983) *Labour-based Construction Programs: A Practical Guide for Planning and Management*, Oxford University Press, New Delhi.

World Bank (1984) *The Construction Industry: Issues and Strategies in Developing Countries*, The World Bank, Washington D.C.

Zeisal J. (1981) *Inquiry by Design*, 1984 edn Cambridge University Press, London.

Subject Index

Author Index